Laboratory Animal Management

RODENTS

Committee on Rodents
Institute of Laboratory Animal Resources
Commission on Life Sciences
National Research Council

NATIONAL ACADEMY PRESS
Washington, D.C. 1996

National Academy Press • 2101 Constitution Avenue, N.W. • Washington, D.C. 20418

NOTICE: The project that is the subject of this report was approved by the Governing Board of the National Research Council, whose members are drawn from the councils of the National Academy of Sciences, National Academy of Engineering, and Institute of Medicine. The members of the committee responsible for the report were chosen for their special competences and with regard for appropriate balance.

This report has been reviewed by a group other than the authors according to procedures approved by a Report Review Committee consisting of members of the National Academy of Sciences, National Academy of Engineering, and Institute of Medicine.

This study was supported by the U.S. Department of Health and Human Services (DHHS) through contract number NO1-CM-07316 with the Division of Cancer Treatment, National Cancer Institute; the Animal Welfare Information Center, National Agricultural Library, U.S. Department of Agriculture (USDA), through grant number 5932U4-8-59; and Howard Hughes Medical Institute through grant number 70209-500104. Additional support was provided by Charles River Laboratories, Wilmington, Massachusetts; Harlan Sprague Dawley, Indianapolis, Indiana; and the following members of the Pharmaceutical Manufacturers Association: Abbott Laboratories, Abbott Park, Illinois; Amgen, Inc., Thousand Oaks, California; Berlex Laboratories, Inc., Cedar Knolls, New Jersey; Bristol-Myers Squibb Co., New York, New York; Bristol-Myers Squibb Pharmaceutical Research Institute, Princeton, New Jersey; Burroughs Wellcome Co., Research Triangle Park, North Carolina; Ciba-Geigy, Summit, New Jersey; Dupont Merck Research & Development, Wilmington, Delaware; Johnson & Johnson, New Brunswick, New Jersey; Marion Merrell Dow Inc., Kansas City, Missouri; Pfizer Inc., Groton, Connecticut; Sandoz Research Institute, East Hanover, New Jersey; Schering-Plough Research, Bloomfield, New Jersey; SmithKline Beecham Pharmaceuticals, King of Prussia, Pennsylvania; Syntex Discovery Research, Palo Alto, California; 3M Corporation, St. Paul, Minnesota; and Wyeth-Ayerst Research, Philadelphia, Pennsylvania.

Core support is provided to the Institute of Laboratory Animal Resources by the Comparative Medicine Program, National Center for Research Resources, National Institutes of Health, through grant 5P40RR0137; the National Science Foundation through grant BIR-9024967; the U.S. Army Medical Research and Development Command, which serves as the lead agency for combined U.S. Department of Defense funding also received from the Human Systems Division of the U.S. Air Force Systems Command, Armed Forces Radiobiology Research Institute, Uniformed Services University of the Health Sciences, and U.S. Naval Medical Research and Development Command, through grant DAMD17-93-J-3016; and research project grant RC-1-34 from the American Cancer Society.

Any opinions, findings, and conclusions or recommendations expressed in this publication are those of the committee and do not necessarily reflect the views of DHHS, USDA, or other sponsors, nor does the mention of trade names, commercial products, or organizations imply endorsement by the U.S. government or other sponsors.

Library of Congress Cataloging-in-Publication Data

Rodents / Committee on Rodents, Institute of Laboratory Animal Resources, Commission on Life Sciences, National Research Council.

p. cm. — (Laboratory animal management series)

"February 1996."

Includes bibliographical references and index.

ISBN 0-309-04936-9

1. Rodents as laboratory animals. I. Institue of laboratory Animal Resources (U.S.). Committee on Rodents. II. Series. SF407.R6R62 1996

619'.93—dc20 96-4532

COMMITTEE ON RODENTS

Bonnie J. Mills (*Chairman*), Biotech Group, Immunotherapy Division, Baxter Healthcare Corp., Irvine, California
Anton M. Allen, Laboratory Animal Health Services Division, Microbiological Associates, Inc., Rockville, Maryland
Lauretta W. Gerrity, Animal Resource Program, University of Alabama at Birmingham, Birmingham, Alabama
Joseph J. Knapka, Veterinary Resources Program, National Center for Research Resources, National Institutes of Health, Bethesda, Maryland
Arthur A. Like, Department of Pathology, University of Massachusetts, Worcester, Massachusetts
Frank Lilly, Department of Molecular Genetics, Albert Einstein College of Medicine, Bronx, New York
George M. Martin, Department of Pathology, University of Washington, Seattle, Washington
Gwendolyn Y. McCormick, Laboratory Animal Resources, Searle, Skokie, Illinois
Larry E. Mobraaten, The Jackson Laboratory, Bar Harbor, Maine
William J. White, Professional Services, Charles River Laboratories, Wilmington, Massachusetts
Norman S. Wolf, Department of Pathology, University of Washington, Seattle, Washington

CONTRIBUTORS

Wallace D. Dawson, Department of Biological Sciences, University of South Carolina, Columbia, South Carolina
Edward H. Leiter, The Jackson Laboratory, Bar Harbor, Maine
Barbara McKnight, Department of Biostatistics, School of Public Health, University of Washington, Seattle, Washington
Glenn M. Monastersky, Transgenics, Charles River Laboratories, Wilmington, Massachusetts
Richard J. Traystman, Department of Anesthesiology and Critical Care Medicine, The Johns Hopkins Hospital, Baltimore, Maryland

Staff

Dorothy D. Greenhouse, Senior Program Officer
Amanda E. Hull, Program Assistant
Norman Grossblatt, Editor

INSTITUTE OF LABORATORY ANIMAL RESOURCES COUNCIL

The Institute of Laboratory Animal Resources (ILAR) was founded in 1952 under the auspices of the National Research Council. A component of the Commission on Life Sciences, ILAR develops guidelines and positions and disseminates information on the scientific, technological, and ethical use of laboratory animals and related biological resources. ILAR promotes high-quality, humane care of laboratory animals and the appropriate use of laboratory animals and alternatives in research, testing, and education. ILAR serves as an advisor to the federal government, the biomedical research community, and the public.

COMMISSION ON LIFE SCIENCES

The National Academy of Sciences is a private, nonprofit, self-perpetuating society of distinguished scholars engaged in scientific and engineering research, dedicated to the furtherance of science and technology and to their use for the general welfare. Upon the authority of the charter granted to it by the Congress in 1863, the Academy has a mandate that requires it to advise the federal government on scientific and technical matters. Dr. Bruce M. Alberts is president of the National Academy of Sciences.

The National Academy of Engineering was established in 1964, under the charter of the National Academy of Sciences, as a parallel organization of outstanding engineers. It is autonomous in its administration and in the selection of its members, sharing with the National Academy of Sciences the responsibility for advising the federal government. The National Academy of Engineering also sponsors engineering programs aimed at meeting national needs, encourages education and research, and recognizes the superior achievements of engineers. Dr. Harold Liebowitz is president of the National Academy of Engineering.

The Institute of Medicine was established in 1970 by the National Academy of Sciences to secure the services of eminent members of appropriate professions in the examination of policy matters pertaining to the health of the public. The Institute acts under the responsibility given to the National Academy of Sciences by its congressional charter to be an adviser to the federal government and upon its own initiative to identify issues of medical care, research, and education. Dr. Kenneth I. Shine is president of the Institute of Medicine.

The National Research Council was established by the National Academy of Sciences in 1916 to associate the broad community of science and technology with the Academy's purposes of furthering knowledge and advising the federal government. Functioning in accordance with general policies determined by the Academy, the Council has become the principal operating agency of both the National Academy of Sciences and National Academy of Engineering in the conduct of their services to the government, the public, and the scientific and engineering communities. The Council is administered jointly by both Academies and the Institute of Medicine. Dr. Bruce M. Alberts and Dr. Harold Liebowitz are chairman and vice-chairman, respectively, of the National Research Council.

Preface

Biomedical and behavioral research, product testing, and many aspects of science education rely heavily on the use of animals. Quality care of these animals is essential, not only for the animals' welfare, but also for obtaining valid data. Environmental and biologic factors can influence experimental results by exerting subtle influences on an animal's physiologic characteristics, behavior, or both. Although there is a tendency to feel more concern for species to which humans develop an attachment (e.g., dogs and cats) and species that are biologically "closer" to humans (nonhuman primates), the same attention to environmental control for and good care of every laboratory species is necessary to ensure the high quality of both science and ethical practice.

Rodents are, by far, the largest group of animals used in research and testing. In 1986, the Office of Technology Assessment estimated that 17-22 million animals were being used each year in the United States, of which about 13.2-16.2 million were rodents (*Alternatives to Animal Use in Research, Testing, and Education*; Pub. No. OTA-BA-273; U.S. Congress Office of Technology Assessment; Washington, D.C.; 1986). In the 15 years since the last Institute of Laboratory Animal Resources report on the general management of rodents was published, important advances in biomedical research and increased public awareness have created a new environment for animal research. Modern technology—such as insertion of functional genes from other species into mice or rats, elimination of a single selected gene or function in mice, and the re-creation of elements of the human immune system in mice—has greatly expanded the usefulness of rodents in drug development and as models of human diseases. The technologic requirements of such advanced systems have led to improved understanding and implementation of environmental requirements for the care and use of rodents in research.

The intent of this report is to provide current information to laboratory animal scientists (including both animal-care technicians and veterinarians), investigators, research technicians, and administrators on general elements of rodent care and use that should be considered both for optimal design and conduct of research and to meet current standards of care and use. We emphasize that this report provides guidelines and should not be used as a substitute for good professional judgment, which is essential in the application of the guidelines. Where possible, we refer to other documents that provide more detail on specific aspects of rodent care and use.

Bonnie J. Mills, *Chairman*
Committee on Rodents

Contents

Laboratory Animal Management

RODENTS

1

Laboratory Animals and Public Perspective

REGULATORY ISSUES

In recent years, virtually every aspect of biomedical research has been increasingly subjected to public scrutiny. A major concern is the justification of public funding. In addition, heightened public awareness and pressure have resulted in increased oversight in such areas as the health and safety of workers, the state of the environment, and the welfare of animals used in research, teaching, and testing. Design and review of protocols involving the use of animals should include consideration of applicable regulations and public accountability in each of those areas.

Two federal laws govern the use of animals. The Health Research Extension Act (PL 99-158), passed in 1985, amended Title 42, Section 289d, of the U.S. Code and gave the force of law to the *Public Health Service Policy on Humane Care and Use of Laboratory Animals* (PHS, 1996; hereafter called *PHS Policy*). *PHS Policy* applies to all activities conducted or funded by the Public Health Service (PHS) that involve any live vertebrate animal used or intended for use in research, training, or testing. It requires compliance with the Animal Welfare Regulations (AWRs), and it specifies minimal components of an institution's animal care and use program, oversight responsibilities, and reporting requirements. Programs for animal care and use must be based on the *Guide for the Care and Use of Laboratory Animals* (NRC, 1996 et seq.), hereafter called the *Guide*; any departure from its recommendations must be documented and justified. *PHS*

Policy stresses institutional self-regulation and gives responsibility for oversight to an institutional animal care and use committee (IACUC). The Office for Protection from Research Risks (OPRR) is responsible for the general administration and coordination of *PHS Policy*. OPRR responsibilities include reviewing and approving (or disapproving) institutional assurances, communicating with institutions concerning implementation of *PHS Policy*, investigating allegations of noncompliance by PHS-funded institutions, reviewing and approving (or disapproving) waivers to *PHS Policy*, and making site visits to selected institutions.

Title 7, Sections 2131 et seq., of the U.S. Code, popularly called the Animal Welfare Act and most recently amended in 1985 by PL 99-198, was originally written in 1966 to protect pets. Its focus has since shifted to protecting laboratory animals. In addition to requiring that the U.S. Department of Agriculture (USDA) establish minimal standards for animal husbandry, care, treatment, and transportation, the act now includes provisions to reduce animal use by eliminating unnecessary duplication and mandates consideration of alternatives to procedures that are likely to cause pain or distress in live animals. The amended act applies to most warm-blooded animals used or intended for use in research, teaching, or testing in the United States. Like *PHS Policy*, it emphasizes institutional self-regulation and gives oversight responsibility to an IACUC. Regulatory Enforcement and Animal Care (REAC), a part of the USDA Animal and Plant Health Inspection Service, administers and enforces the regulations (9 CFR 1-3) and carries out inspections of facilities to determine compliance. Laboratory mice (genus *Mus*) and rats (genus *Rattus*), which make up more than 90 percent of the animals used in research, are not covered by the AWRs and are not subject to REAC inspection. However, there is a movement to include them; the decision on this issue is likely to be made in federal court.

Other regulations, policies, and guidelines address animal-care issues, although they are not specifically directed at animal research. They include the Good Laboratory Practice rules promulgated by the Food and Drug Administration (21 CFR 58) and the Environmental Protection Agency (40 CFR 160 and 40 CFR 792), which provide standards for the care and housing of test animals, and *Biosafety in Microbiological and Biomedical Laboratories* (Richmond and McKinney, 1993), which provides guidelines for containment of animals and animal wastes during and resulting from animal experimentation with pathogens.

For reviews and discussions of the various regulations and guidelines, refer to *Education and Training in the Care and Use of Laboratory Animals: A Guide for Developing Institutional Programs*, Part III, Chapter 1 (NRC, 1991); *Use of Laboratory Animals in Biomedical and Behavorial Research*, Chapter 5 (NRC, 1988); *The Biomedical Investigator's Handbook for Researchers Using Animal Models*, Chapter 6 (Foundation for Biomedical Research, 1987); and *The Institutional Animal Care and Use Committee Guidebook (IACUC Guidebook)* (ARENA/OPRR, 1992).

In addition to the regulations noted above, animal experimentation with hazardous agents is subject to regulations that govern handling, use, and disposal of hazardous agents, such as radioisotopes and toxic chemicals. Likewise, protection of workers from a variety of potential workplace hazards is mandated by occupational safety and health agencies at the federal level and, in many cases, at the state level. It is the responsibility of each investigator using animals to know and comply with relevant regulations, guidelines, and policies (federal, state, local, and institutional).

ETHICAL CONSIDERATIONS

The laws, regulations, policies, and guidelines discussed above establish common standards for the humane care and use of laboratory animals. Recent revisions have refined earlier standards and improved the well-being of laboratory animals. Nevertheless, it is the obligation of every investigator who uses animals to ensure that the highest principles of humane care and use are applied. These principles are summarized in the U.S. government "Principles for the Utilization and Care of Vertebrate Animals Used in Testing, Research, and Training" (published in NRC, 1996, pp. 116-118, and PHS, 1996, p. 1), which was prepared by the Interagency Research Animal Committee, a group whose main concerns are the conservation, use, care, and welfare of research animals. The principles address such issues as the value of the proposed work; selection of appropriate models; minimization of pain and distress; use of sedation, analgesia, or anesthesia when painful procedures are necessary; euthanasia of animals that might suffer severe or chronic pain or distress; provision of appropriate housing and veterinary care; training of personnel; and IACUC oversight of exceptions to the principles. The principles emphasize the role of the IACUC in determining the appropriateness and value of proposed work in which animals are likely to be subjected to unalleviated pain or discomfort. Some kinds of research should be especially carefully reviewed and periodically re-evaluated by IACUCs, including studies that involve unalleviated pain or distress (such as those in which death is the end point) and studies that involve food or water deprivation.

Some people and groups question the value of using animals in biomedical research and suggest that the knowledge gained is not sufficiently applicable to human disease to justify the pain, distress, and loss of life suffered by laboratory animals. However, Nicoll and Russell (1991) point out that animal research has contributed in an important way to 74 percent of 386 major biomedical advances made from 1901 to 1975 and that 71 percent of the 82 Nobel prizes for physiology or medicine awarded from

1901 to 1982 were given for research that depended on studies with animals. The regular occurrence of new infectious diseases of humans and animals—such as Legionnaire's disease, AIDS, Lyme disease, and canine parvovirus disease—and the existence of diseases that kill hundreds of thousands of people and animals a year—such as cancer, cardiovascular disease, and stroke—make research in living systems imperative if we wish to continue to make medical progress.

Most of the public are rightly concerned with the elimination of unnecessary animal suffering and the protection of pets, and it is an obligation of scientists to educate the press, the legislature, and the public about the efforts made by the scientific community to minimize animal pain and suffering, the extensive review to which animal research is subjected, and the great benefits we and our pets derive from animal research. These benefits include the development of antiviral vaccines (e.g., vaccines against poliovirus, canine parvovirus, and feline leukemia virus), advances in tissue transplantation (e.g., of kidneys, corneas, skin, heart, liver, and bone marrow), and the development of new treatments for cardiovascular disease (e.g., open-heart surgery, valve replacement, and artery replacement). The educational process should stress that scientists and most of the public agree that the use of animals in research is necessary, that animals should be cared for and used as humanely as possible, and that unnecessary suffering should be prevented. Results of such educational efforts are beginning to appear in the form of state and federal legislation to protect animal-research facilities and laboratories from vandalism. The educational process should continue, and all scientists should be committed to it.

Useful discussions of the ethical issues related to animal research can be found in *Use of Laboratory Animals in Biomedical and Behavioral Research* (NRC, 1988); *The Biomedical Investigator's Handbook for Researchers Using Animal Models* (Foundation for Biomedical Research, 1987); *Mozart, Alexander the Great, and the Animal Rights/Liberation Philosophy* (Nicoll and Russell, 1991); and *Education and Training in the Care and Use of Laboratory Animals: A Guide for Developing Institutional Programs*, Part III, Chapter 2 (NRC, 1991).

REFERENCES

ARENA/OPRR (Applied Research Ethics National Association and Office for Protection from Research Risks). 1992. Institutional Animal Care and Use Committee Guidebook. NIH Pub. No. 92-3415. Washington, D.C.: U.S. Department of Health and Human Services. Available from either ARENA, 132 Boylston Street, Boston, MA 02116 or U.S. Government Printing Office, Washington, DC 20402 (refer to stock no. 017-040-00520-2).

Foundation for Biomedical Research. 1987. The Biomedical Investigator's Handbook for Researchers Using Animal Models. Washington, D.C.: Foundation for Biomedical Research. 86 pp.

Nicoll, C. S., and S. M. Russell. 1991. Mozart, Alexander the Great, and the animal rights/liberation philosophy. FASEB J. 5:2888-2892.

NRC (National Research Council), Institute of Laboratory Animal Resources Committee to Revise the Guide for the Care and Use of Laboratory Animals. 1996. Guide for the Care and Use of Laboratory Animals, 7th edition. Washington, D.C.: National Academy Press.

NRC (National Research Council), Commission on Life Sciences and Institute of Medicine, Committee on the Use of Laboratory Animals in Biomedical and Behavioral Research. 1988. Use of Laboratory Animals in Biomedical and Behavioral Research. Washington, D.C.: National Academy Press. 102 pp.

NRC (National Research Council), Institute of Laboratory Animal Resources, Committee on Educational Programs in Laboratory Animal Science. 1991. Education and Training in the Care and Use of Laboratory Animals: A Guide for Developing Institutional Programs. Washington, D.C.: National Academy Press. 139 pp.

PHS (Public Health Service). 1996. Public Health Service Policy on Humane Care and Use of Laboratory Animals. Washington, D.C.: U.S. Departmentof Health and Human Services. 16 pp. Available from the Office for Protection from Research Risks, National Institutes of Health, 6100 Executive Boulevard, MSC 7507, Suite 3B01, Rockville, MD 20892-7507.

Richmond, J. Y., and R. W. McKinney, eds. 1993. Biosafety in Microbiological and Biomedical Laboratories, 3rd ed. HHS Pub. No. (CDC) 93-8395. Washington, D.C.: U.S. Department of Health and Human Services. Available from Superintendent of Documents, U.S. Government Printing Office, Washington, DC 20402.

2

Responsibilities of Animal Care and Use Committees

PROGRAM OVERSIGHT

The Animal Welfare Regulations, or AWRs (9 CFR 2.31), mandate that each institution in which warm-blooded animals other than birds, rodents of the genera *Mus* and *Rattus*, and farm animals are used in research, testing, or education have an institutional animal care and use committee (IACUC) to oversee the institution's animal care and use program. *Public Health Service Policy on Humane Care and Use of Laboratory Animals*, or *PHS Policy* (PHS, 1996), has the same requirement for each PHS-funded institution that uses live vertebrates. Program oversight is more than semiannual facility inspections and protocol reviews; it places a more global responsibility on the IACUC for general oversight of the animal program. In a quality program, the highest standards of science and ethics are understood and supported at every level of animal use, from the animal-care technician to the program administrator.

Program oversight should include consideration of all institutional functions, policies, or practices that directly affect the care and use of laboratory animals. It might include training; occupational health and safety; the veterinary-care program; use of animals in teaching; consistency of institutional policies with local, state, and federal regulations; interactions with other internal groups, such as those responsible for space allocation, research administration, and biosafety; interactions with external groups, such as funding agencies; specific concerns or complaints about animal use; investigation of unauthorized activities involving the use of animals; and effective communication between investigators, animal-care staff, and administrators.

An IACUC customarily reviews programs at the same time that it conducts semiannual facility inspections. It is important to document that both the program and the facilities have been reviewed by the IACUC and to note program improvements, as well as program deficiencies. Results of semiannual reviews must be provided to the institutional official and must include a plan for correcting deficiencies and minority views (9 CFR 2.31c3; 9 CFR 2.35a3; PHS, 1986).

PROTOCOL REVIEW

One of the many important responsibilities of an IACUC is to review the protocols for research, testing, or teaching projects in which any species covered by the AWRs or *PHS Policy* will be used. The protocol-review mechanism is designed to ensure that investigators consider the care and use of their animals and that procedures comply with federal, state, and institutional regulations and policies. In addition, the review mechanism enables an IACUC to become an important institutional resource, assisting investigators in all areas involving the use of animals.

Each research protocol should include the following information, much of which is required by the AWRs, *PHS Policy*, or both:

- the purpose of the study;
- the rationale for selection of the species and the numbers of animals to be used;
- the strain, sex, and age of the animals to be used;
- the living conditions of the animals, particularly special housing and husbandry requirements;
- the experimental methods and manipulations;
- justification of multiple major survival surgeries on any individual animal;
- preprocedural and postprocedural care and medications;
- procedures that will be undertaken to avoid or minimize more than momentary discomfort, pain, and distress, including, where appropriate, the use of anesthetics, analgesics, and tranquilizers;
- if experimental manipulation is likely to cause more than momentary or slight pain or distress that for scientifically valid reasons cannot be relieved by appropriate drugs, the process undertaken to ensure that there are no appropriate alternatives (some types of research, such as trauma studies and studies in which death is the end point, are particularly sensitive in this regard);

- procedures that will be used to monitor the animals in studies in which close monitoring is required, for example, those involving food or water deprivation and tumor growth (studies that require close monitoring should include specific end points);
- procedures and justification for long-term restraint;
- the euthanasia method, including a justification if it is not consistent with the recommendations of the American Veterinary Medical Association Panel on Euthanasia (AVMA, 1993 et seq.);
- assurance that the protocol does not unnecessarily duplicate previous work; and
- the qualifications of personnel who will perform the procedures outlined.

Protocol submission and review formats differ widely from one institution to another and depend on a number of variables, including the size and mission of the institution, other levels of scientific review to which the protocol will be subjected, and past experiences of the IACUC. Thorough and careful preparation of a protocol will facilitate the review process and reduce delay. One review approach used by IACUCs, particularly in large institutions, is to assign a knowledgeable committee member to each protocol as the primary reviewer. The primary reviewer deals directly with the investigator to clarify issues in question. Changes or clarifications in the protocol that result from the reviewer's discussions with the investigator are submitted to the IACUC in writing. Later, at an IACUC meeting, the primary reviewer presents and discusses the protocol and relates discussions with the investigator. After the reviewer's presentation of the protocol, the reviewer recommends a course of action, which is then discussed and voted on by the IACUC. Another kind of protocol review (which is especially effective in small institutions with few protocols) is initial review by the entire IACUC. Many committees rely on additional review by experts (either on or outside the committee) in specific subjects; for example, a veterinarian should review protocols for appropriateness of the proposed anesthesia and analgesia, and a statistician might review statistically complicated study designs. In some institutions, such as pharmaceutical companies, some kinds of studies (e.g., pharmaceutical development and toxicology screening) are based on standard operating procedures. Nevertheless, IACUC review and approval are required before study initiation.

Several outcomes of protocol review are possible: approval, approval contingent on receipt of additional information (to respond to minor problems with the protocol), deferral and rereview after receipt of additional information (to respond to major problems with the protocol), and withholding of approval. If approval of a protocol is withheld, an investigator should be given the opportunity to respond to the critique of the IACUC in writing, to appear in person at an IACUC meeting to present his or her viewpoint, or both. It is also important that expedited review be possible; however, the use of expedited review does not negate the requirement (9 CFR 2.31; PHS, 1996, Section IV.C.2) that each IACUC member be given the opportunity to review every protocol and to call for a full committee review before approval is given (McCarthy and Miller, 1990).

The question of protocol review for scientific merit has been handled in a variety of ways by IACUCs. Many protocols are subjected to extensive, external scientific review as part of the funding process; in such instances, the IACUC can be relatively assured of appropriate scientific review. For studies that will not undergo outside review for scientific merit, many IACUCs require signoff by the investigators, department chairmen, or internal review committees; this makes signers responsible for providing assurance that the proposed studies have been designed and will be performed "with due consideration of their relevance to human or animal health, the advancement of knowledge, or the good of society" (NRC, 1996, p.116; PHS, 1996, p.1). Occasionally, IACUC members and scientists differ as to the relevance of proposed studies to human and animal health and the advancement of knowledge. Each institution should develop guidelines for dealing with this potential conflict.

It is important that the IACUC document the protocol-review process, so that it is clear that all aspects of a project, especially aspects that might seriously affect animal well-being, have been thoroughly considered by the IACUC; minority views must be included (9 CFR 2.31). IACUCs should keep accurate records, pay careful attention to semantics, and be familiar with local, state, and federal "freedom of information" laws that make records available to the general public on request.

PERSONNEL QUALIFICATIONS AND TRAINING

Job applicants for positions that require access to an animal facility should be carefully screened. Checks for records of criminal activity might be warranted. Potential employees should understand clearly the nature of the work. Education of animal-care and research personnel regarding proper security procedures is critical to ensuring facility security. This training should be part of new-employee orientation and should be reinforced frequently.

Both *PHS Policy* (PHS, 1996) and the AWRs (9 CFR 3.32) require that institutions provide training on the care and use of animals. It is the responsibility of the IACUC to ensure that animal-care and research staff are appropriately trained (PHS, 1996). As part of program oversight, the IACUC must ensure that procedures for providing and documenting training are in place; however, the responsibility for design and implementation of training programs varies. Responsibilty for course objectives and format is frequently shared by staff from various functional units, such as veterinary staff, employee-health personnel, safety officers, and IACUC members.

People for whom it is required that training be made available (9 CFR 2.32) include those who provide animal husbandry (caretakers), those who perform technical procedures on animals (research staff and animal technicians and technologists), those who provide veterinary medical care and treatment (veterinarians and veterinary technicians). The National Research Council has recommended that training also be provided to other personnel, including administrative and housekeeping staffs. Training is also important for those who are responsible for oversight (IACUC members and administrators). The varied backgrounds and responsibilities of the people for whom training is provided, the size and nature of the institution, the variety and numbers of animals used, and the nature of animal use (i.e., research, teaching, and testing) are important in the design of an institutional training program. The program should be tailored to meet the institution's specific needs and designed with ease of use and convenience in mind. Although the format and content might vary considerably between institutions, there is some agreement on minimal information that should be provided. The following topics are considered by the National Research Coucil to be essential elements of a basic training program (NRC, 1991):

- laws, regulations, and policies that affect the care and use of animals;
- ethical and scientific issues;
- alternatives to the use of animals;
- responsibilities of the IACUC and the research and veterinary staffs;
- pain and distress;
- anesthetics, analgesics, tranquilizers, and neuromuscular blocking agents;
- survival surgery and postsurgical care;
- euthanasia;
- husbandry, care, and the importance of the environment; and
- resources for additional information.

For each of those elements, all personnel should be provided a general overview that is designed to promote understanding of and facilitate compliance with regulations and policies. Depending on the audience and the topic, it might not be necessary to provide a high degree of detail. For example, the discussion of survival surgery should familiarize the audience with regulations and acceptable standards for surgical procedures and postsurgical care, but it need not provide details of specific surgical methods, which would be important only to those performing or assisting with the surgery or postsurgical care.

In contrast, substantial detail should be provided to people in direct contact with animals, and the content should be appropriate to their responsibilities for animal care or use. For example, detailed information on species-specific housing methods, husbandry procedures, and handling techniques should be provided to animal caretakers; research staff should be specifically qualified through training or experience for each approved procedure in the designated species; and veterinary staff should be appropriately trained in relevant aspects of laboratory animal medicine.

Training is provided in various ways. Many people are qualified in animal care, use, or specific procedures by having formal training in degree or certification programs (e.g., veterinarians certified in laboratory animal medicine, certified animal technologists and technicians, and physicians with surgical specialties). Others might be qualified by having previous experience (e.g., investigators who have research experience with a particular animal model). Regardless of the extent of previous training, it is wise for each institution to provide information about the standards, requirements, and expectations of the institution and an updated overview of key issues to all personnel involved with animal care or use.

Institutions often need to provide extensive training to staff that provide daily care and observation of animals or to research personnel without previous or recent experience in a particular technique or species. Various methods can be used, including lectures and seminars, videotaped lectures and demonstrations, and observation by experienced personnel. Continuing-education courses are available in many areas, particularly at or near large institutions or universities, and attendance can be encouraged by tuition-reimbursement programs. Each method has advantages and disadvantages, and each institution should select the format that serves the needs of its staff best.

Resources for developing training programs include qualified institutional staff, formal courses by recognized organizations (e.g., the American Association for Laboratory Animal Science), and written and audiovisual training aids (see NRC, 1991, part IV, chapter 3).

It is important not only to ensure or provide appropriate training, but to document that all personnel who care for or use animals are appropriately trained. Training and education can be documented in a variety of ways. For example, previous training can be documented by records, publications, and signed statements of experience, and training provided by the institution can be documented by attendance records, signed statements, and notes to personnel files. A powerful method for documenting or monitoring the qualifications of personnel is observation of animal procedures by a qualified person. This method provides an accurate assessment of the expertise of the person performing the procedure, as well as information about the health status of the animal during the procedure. Such observation is usually considered to be an appropriate component of veterinary oversight.

OCCUPATIONAL HEALTH AND SAFETY

An occupational health and safety program is an important component of the operation of any institution in which animals are used (NRC, In press). This program should seek to safeguard the health of employees that work with laboratory animals by developing standard operating procedures to minimize the chance of exposure to zoonotic diseases and providing the necessary training so that employees will understand the risks associated with working with animals and the importance of complying with institutional procedures. The program can also serve the animals being maintained by screening employees for zoonotic diseases and, where appropriate, providing immunizations that will minimize the likelihood of introduction of zoonotic agents into the animal facility.

The design of an occupational health and safety program should be based on a careful review of the potential hazards that exist in the animal facilities. The program must comply with Occupational Safety and Health Administration (OSHA) standards (29 CFR 110-114) and should be designed with the aid of medical personnel who are knowledgeable in occupational medicine and familiar with zoonotic diseases. Each aspect of the program should be carefully and realistically evaluated with respect to the magnitude of risk involved, the legal and practical enforceability of mandated components of the program, and the costs relative to the likelihood of detecting or preventing a problem. A legal review of the final proposed program is advisable because local, state, and federal laws might preclude adoption or enforcement of some of its components.

Oversight of occupational health and safety programs varies among institutions. It is frequently assigned to employee-health staff, but in some institutions it is the reponsibility of personnel, human-resources, veterinary, or other administrative staffs. Generally, an IACUC verifies during its semiannual review that the occupational health and safety program is in place and that its components are appropriate to the institution's animal care and use program.

Few general rules can be applied to occupational health and safety programs for rodent facilities. Only a few rodent diseases pose a threat to humans, and many of these have a very low prevalence (e.g., the diseases caused by Hantaan virus, lymphocytic choriomeningitis virus, some *Salmonella* species, *Hymenolepis nana*, and *Streptobacillus moniliformis*). In most cases, prophylactic immunizations do not exist for rodent zoonotic organisms; if immunizations do exist, the risks associated with them should be balanced against the likelihood of contracting the disease. Personnel should be instructed to notify their supervisors of bite wounds, unusual illnesses, and suspected health hazards. Facilities often maintain records of individual work assignments and of employee-reported problems. That information, if kept accurately and evaluated regularly, can be of value in alerting both the institution and employees to unusual patterns of illness that could indicate an animal-related disease.

Other occupational hazards, including allergies, should be recognized, and methods should be developed for minimizing the risks and treating problems if they occur. Animal-care personnel are generally at greater risk of contracting tetanus than other segments of the workforce because the greater frequency with which they handle animals puts them at greater risk of being bitten. Therefore, it is important that immunization against tetanus be offered to animal-care personnel and that a record of prophylactic immunizations be kept.

Exposure to potentially toxic materials and ergonomic practices associated with lifting and moving equipment and materials are also of concern in rodent facilities. The animal facilities and related support areas should be evaluated for the need for protective devices (e.g., respirators, lifting-support belts and gloves, and ear and eye protection) and for the need to develop safety measures peculiar to the tasks being conducted. If animal-care, research, and maintenance personnel could be exposed to potentially hazardous biologic, chemical, or physical agents, the exposure to such agents should be monitored. Specific safety procedures designed to minimize the risk of exposure should be developed in consultation with appropriate health and safety professionals.

The gathering of pre-employment health information—by questionnaire, physical examination conducted by a physician, or both—might be deemed appropriate, provided that such information is related specifically to evaluating the employee's potential for carrying zoonotic organisms or having predisposing conditions (e.g., allergies, immunosuppression, pregnancy, and heart disease) that would make exposure to animals hazardous to his or her health. All medical records must be kept confidential, should be reviewed by a competent health care professional, and must not be used to gather information on non-animal-related health matters that could be used to prevent hiring the employee. Conditions identified that might affect the animal care and use program (e.g., a positive result of a test for tuberculosis) or might put an employee at increased risk (e.g., pregnancy) should be communicated to appropriate personnel to minimize unnecessary risk to employees, animals, or both. The conditions for employment and use of employee-health information should be precisely defined in advance by the institution and should comply with local, state, and federal requirements.

Periodic physical examinations might be offered to some employees in some job categories. In some institutions, programs have also been established to obtain and store individual serum samples taken before hiring and during employment for future diagnostic purposes. In general, such serum-banking procedures are seldom undertaken in rodent facilities and, when offered, are usually voluntary. In institutions in which research involving the use of zoonotic agents in rodents is conducted and in which there is a substantial risk of infection, prophylactic vaccinations, if available, should be offered to employees at risk; in such cases, it is important that employees be informed by trained medical personnel of both the benefits and the risks associated with the vaccinations.

An important component of the occupational health and safety program is employee education. Each institution should have in place a course of study consisting of lectures or seminars, self-help materials, or both to instruct personnel who work with animals about zoonoses, allergies to animals, the importance of personal hygiene, special risks associated with pregnancy, and other appropriate topics. This course of study should also include information on hazardous materials that are used in the facilities, including those regulated by the Environmental Protection Agency and the Nuclear Regulatory Commission and those used in procedures evaluated by OSHA. Of particular importance are chemical agents used in routine animal-care operations, including disinfectants, cage-cleaning solutions, and sterilizing agents.

USE OF HAZARDOUS AGENTS

Biomedical experimentation frequently involves the use of hazardous agents, which can be classified as chemical (e.g., chemical carcinogens and chemotherapy agents), physical (e.g., radioisotopes), or biologic (e.g., infectious agents and recombinant DNA). In addition to the common concerns associated with handling and storage, the use of these agents in animals introduces unique concerns, including hazards associated with administration of the agents to the animals, the mode and quantity of excretion of the agents by the animals, contact with contaminated animal tissues, and disposal of carcasses, bedding, and excrement.

It is the responsibility of the IACUC to ensure that the procedures for use and monitoring of hazardous agents have been reviewed and are appropriate (NRC, 1996 et seq.). That is commonly and most readily accomplished by requiring that any use of hazardous agents be approved by an appropriate institutional safety committee (e.g., radiation-safety committee, infectious-agents committee, biosafety committee, or recombinant-DNA-use committee) before IACUC consideration. Formal programs should be in place to review the procedures, facilities, and staff competence for the proposed studies and to monitor compliance with federal, state, and local regulations and institutional policies during the conduct of the research. Requirements of both hazard containment and good animal husbandry should be met. Areas in which hazardous agents are approved for use should be visited as part of the IACUC semiannual inspection. Review should include assurance that there are universal warning signs where hazardous agents are contained and used and that all involved personnel are familiar with and are using approved procedures.

In addition to hazardous agents for which regulations or guidelines are well established—such as radioisotopes (10 CFR 20), infectious agents (NCI, 1974; NIH, 1984; Richmond and McKinney, 1993), and human-blood products (29 CFR 1910)—it is important that there be equal oversight of the use of experimental agents not usually thought of as hazardous, such as some categories of agents for human therapy, fresh tissue from humans or animals, cultured cell lines that might harbor pathogens, and volatile anesthetics. A list of publications pertaining to regulations and guidelines for the use of hazardous agents can be found in the *Guide* (NRC, 1996 et seq.).

REFERENCES

AVMA (American Veterinary Medical Association). 1993. 1993 Report of the AVMA Panel on Euthanasia. J. Am. Vet. Med. Assoc. 202:229-249.

McCarthy, C. R., and J. G. Miller. 1990. OPRR Reports, May 21, 1990. Available from Office for Protection from Research Risks (OPRR), National Institutes of Health, 6100 Executive Boulevard, MSC 7507, Rockville, MD 20892-7507.

NCI (National Cancer Institute). 1974. Safety Standards for Research Involving Oncogenic Viruses. DHEW Pub. No. (NIH) 75-790. Washington, D.C.: U.S. Department of Health, Education and Welfare. 20 pp.

NIH (National Institutes of Health). 1984. Guidelines for Research Involoving Recombinant DNA Molecules. Fed. Regist. 49(227):46266-46291.

NRC (National Research Council), Institute of Laboratory Animal Resources, Committee to Revise the Guide for the Care and Use of Laboratory Animals. 1996. Guide for the Care and Use of Laboratory Animals, 7th edition. Washington, D.C.: National Academy Press

NRC (National Research Council), Institute of Laboratory Animal Resources, Committee on Educational Programs in Laboratory Animal Science. 1991. Education and Training in the Care and Use of Laboratory Animals: A Guide for Developing Institutional Programs. Washington, D.C.: National Academy Press. 139 pp.

NRC (National Research Council), Institute of Laboratory Animal Resources, Committee on Occupational Safety and Health in Research Animal Facilities. Occupational Health and Safety in the Care and Use of Research Animals. Washington, D.C.: National Academy Press.

PHS (Public Health Service). 1996. Public Health Service Policy on Humane Care and Use of Laboratory Animals. Washington, D.C.: U.S. Departmentof Health and Human Services. 16 pp. Available fromthe Office for Protection from Research Risks, National Institutes of Health, 6100 Executive Boulevard, MSC 7507, Suite 3B01, Rockville, MD 20892-7507.

Richmond, J. Y., and R. W. McKinney, eds. 1993. Biosafety in Microbiological and Biomedical Laboratories, 3rd ed. HHS Pub. No. (CDC) 93-8395. Washington, D.C.: U.S. Department of Health and Human Services. Available from Superintendent of Documents, U.S. Government Printing Office, Washington, DC 20402.

3

Criteria for Selecting Experimental Animals

SPECIES AND STOCKS

Choosing a Species for Study

For a scientific investigation to have the best chance of yielding useful results, all aspects of the experimental protocol should be carefully planned. If animal models will be used, an important part of the process is to consider whether nonanimal approaches exist. If, after careful deliberation and review of the existing literature, the investigator is satisfied that there are no suitable alternatives to the use of live animals for the study in question, the next question that should be addressed is what species would be most appropriate to use.

In choosing a species for study, it is important to weigh a variety of scientific and operational factors, including the following:

- In which species is the physiologic, metabolic, behavioral, or disease process to be studied most similar to that of humans or other animals to which the results of the studies will be applied?
- Do other species possess biologic or behavioral characteristics that make them more suitable for the planned studies (e.g., generation time and availability)?
- Does a critical review of the scientific literature indicate which species has provided the best, most applicable historical data?

- Do any features of a particular species or strain—including anatomic, physiologic, immunologic, or metabolic characteristics—render it inappropriate for the proposed study?
- In light of the methods to be used in the study, would any physical or behavioral characteristics of a particular species make the required physical manipulation or sampling procedures impossible, subject to unpredictable failure, or difficult to apply?
- Does the proposed study require animals that are highly standardized either genetically or microbiologically?

Those and other considerations often lead to the selection of a laboratory rodent species as the most appropriate model for a biomedical research protocol. Rodents are generally easy to obtain and relatively inexpensive to acquire and maintain. Other advantages of laboratory rodents as research models include small size, short generation time, and availability of microbiologically and genetically defined animals, historical control data, and well-documented information on physiologic, pathologic, and metabolic processes.

The order Rodentia encompasses many species. The most commonly used rodents are laboratory mice[1], laboratory rats (*Rattus norvegicus*), guinea pigs (*Cavia porcellus*), Syrian hamsters (*Mesocricetus auratus*), and gerbils (*Meriones unguiculatus*). All those rodents have been extensively studied in the laboratory, and information about them can be found in the peer-reviewed literature and in a number of texts (e.g., Altman and Katz, 1979a,b; Baker et al., 1979-1980; Foster et al., 1981-1983; Fox et al., 1984; Gill et al., 1989; Harkness and Wagner, 1989; Van Hoosier and McPherson, 1987; Wagner and Manning, 1976).

Rodent Stocks

The same factors used in selecting a species for study can be used in selecting a rodent stock. Rodents have been maintained in the laboratory environment for more than 100 years. Some, such as the mouse, have been very well characterized genetically and have undergone genetic manipulation to produce animals with uniformly heritable phenotypes. A hallmark of good scientific method is reproducibility, which is accomplished by minimizing and controlling extraneous variables that can alter research results. In studies that are mechanistic, genetic uniformity is highly desirable. In contrast, genetic uniformity might be undesirable in studies that explore the diversity

[1]Laboratory mice are neither pure *Mus domesticus* nor pure *Mus musculus*; therefore, geneticists have determined that there is no appropriate scientific name (International Committee on Standardized Genetic Nomenclature for Mice, 1994a).

of application of a phenomenon over a range of phenotypes, such as product-registration studies, including safety evaluation of compounds that have therapeutic potential. In many such studies, a varied genetic background might be appropriate, as long as the range of variation can be characterized and is to some degree reproducible (Gill, 1980).

Genetically Defined Stocks

Inbred Strains. The mating of any related animals will result in inbreeding, but the most common and efficacious method for establishing and maintaining an inbred strain is brother x sister (i.e., full-sib) mating in each generation. Full-sib inbreeding for 20 generations will result in more than 98 percent genetic homogeneity, at which point the members of the stock are isogenic, and the stock is considered an inbred strain. Many inbred strains of mice and rats have been developed (Festing, 1989; Festing and Greenhouse, 1992), and they are widely used in biomedical research. Many of the commonly used strains have been inbred for over 200 generations. A few inbred strains of guinea pigs, Syrian hamsters, and gerbils have also been developed (Altman and Katz, 1979b; Festing, 1993; Hansen et al., 1981).

The isogeneity of the members of an inbred strain provides a powerful research tool. Although some genes might remain heterogeneous, most metabolic or physiologic processes, as well as their phenotypic expression, will be identical among individuals of an inbred strain, thereby eliminating a source of experimental variation. Isogeneity also allows exchange of tissue between individuals of an inbred strain without rejection.

F1 Hybrids. F1 hybrid animals are the first filial generation (the F1 generation) of a cross between two inbred strains. They are often more hardy than animals from either of the parental strains, having what is called hybrid vigor. F1 hybrids are heterozygous at all genetic loci at which the parental strains differ; nevertheless, they are *uniformly* heterozygous. Because of the heterogeneity, F1 hybrids will not breed true; to produce them one must always cross animals of the parental inbred strains. Reciprocal hybrids are developed by reversing the strains from which the dam and the sire are taken. Reciprocal male hybrids will have Y-chromosome differences. Reciprocal female hybrids will have identical genotypes but might have differences caused by inherited maternal effects. F1 hybrids will accept tissue from either parental strain, except in the case of a Y-chromosome incompatibility (e.g., a skin graft from a male of either parental strain will be rejected by a female F1 hybrid).

Special Genetic Stocks. The effects of specific genes or chromosomal regions can be studied by using various breeding or gene manipulation methods to create a new strain that differs from the original strain by as little as a single gene.

- *A segregating inbred strain* is an inbred strain maintained by full-sib matings; however, male-female pairs are selected for mating so that one pair of genes will remain heterozygous from generation to generation. This method of mating permits well-controlled experiments because a single sibship contains both carriers and noncarriers of the gene of interest, and all the animals are essentially identical except for that gene.
- A *coisogenic strain* is an inbred strain in which a single-gene mutation has occurred and has been preserved; it is otherwise identical with the nonmutant parental strain. If the mutation is not deleterious when homozygous, the strain can be maintained by simple full-sib matings. If the mutation adversely affects breeding performance, the coisogenic strain can be maintained by one of several special breeding systems (Green, 1981; NRC, 1989). To avoid subline divergence between the coisogenic strain and the nonmutant parental inbred strain, periodic back-crossing (see next paragraph) with the parental strain is recommended.
- A *congenic strain* is a close approximation to a coisogenic strain. It is created by mating an individual that carries a gene of interest, called the differential gene, with an individual of a standard inbred strain. An offspring that carries the differential gene is mated to another individual of the same inbred strain. This type of mating, called back-crossing, is continued for at least 10 generations to produce a congenic strain. Back-crossing for 10 generations minimizes the number of introduced genes other than the differential gene and its closely linked genes. Details on developing congenic strains have been published (Bailey, 1981; Green, 1981). Both coisogenic and congenic strains can be maintained by full-sib matings if the differential gene is homozygous; however, to avoid subline divergence between the congenic strain and the standard inbred strain, periodic back-crossing with the standard strain is recommended.
- A *transgenic strain* is similar to a coisogenic or congenic strain in that it carries a segment of genetic information not native to the strain or individual (Hogan et al., 1986; Merlino, 1991). The introduced genetic material can be from the same or another species. Transgenic animals are described in more detail in Chapter 8.
- *Recombinant inbred (RI) strains* are sets of inbred strains produced primarily to study genetic linkage. Each RI strain is derived from a cross between two standard inbred strains. Animals from the F1 generation are then bred to produce the second filial generation (the F2 generation), members of which are randomly selected and mated to produce a series of RI lines. Members of the F2 generation are used to found RI lines because, unlike the F1 generation, they are not isogenic. The mice derived from any parental pair will be genetically homogeneous when inbreeding is complete;

however, each line in a set will be homozygous for a given combination of alleles originating from the two parental inbred strains. Alleles that are linked in the parental strains will tend to remain together in the RI lines; this is the basis for their use in genetic-mapping studies.

- *Recombinant congenic strains* are like recombinant inbred strains except that each strain of a series has been derived from a back-cross instead of an F2 cross (Demant, 1986). The number of back-crosses made before full-sib inbreeding is started determines the proportion of genes from each of the parental inbred strains. Series of recombinant congenic strains are particularly useful in the genetic analysis of multiple-gene systems, such as that responsible for cancer susceptibility.

Nongenetically Defined Stocks

The terms *noninbred*, *random-mated*, and *outbred* are all used to refer to populations of animals in which, theoretically, there is no genetic uniformity between individuals. Nongenetically defined stocks make up the majority of rodents used in biomedical research and testing, and they are generally less expensive and more readily available than genetically defined stocks.

Noninbred refers to a population of animals in which no purposeful inbreeding system has been established. *Random-mated* refers to a group of animals in which the selection of breeding animals is random. It assumes an almost infinite population with no external selection pressures. In practice, such a colony probably does not exist. *Outbred* refers to a colony in which breeding is accomplished by a purposeful scheme that minimizes or eliminates inbreeding. Animals produced by these breeding systems have varied genotypes, and characterizing the range and distribution of phenotypes requires a large sample of the population.

The degree of heterozygosity in any nongenetically defined stock is continuously varying, so two populations developed from the same parental stock will show differing degrees of heterozygosity at any loci at any time. Spontaneous mutations can occur and become fixed because no purposeful selection is imposed on the population to eliminate the mutant genes. Outbred populations are always evolving and therefore are more variable than inbred strains. For that reason, large sample numbers are needed to account for phenotypic variation that could have an impact on the charactersitics being studied. If outbred animals are used, treatment and control groups in a study will not necessarily be identical, nor will the population of animals necessarily be identical if the study is repeated. The genetic variation in outbred stocks, which can be magnified by sampling error, can make results from different laboratories difficult to compare. Background data on stock characteristics will vary over time, so concurrent controls are needed to allow useful interpretation of data.

STANDARDIZED NOMENCLATURE FOR RODENTS

Standardized nomenclature allows scientists to communicate briefly and precisely the genetics of their research animals. The International Committee on Standardized Genetic Nomenclature for Mice and the International Rat Genetic Nomenclature Committee, which are affiliated with the International Council for Laboratory Animal Science, are responsible for maintaining the nomenclatures for genetically defined mice and rats, respectively, and modifying them as necessary. The sections below briefly describe the nomenclature for inbred, mutant, and outbred mice and rats. The complete rules for mice can be found in the third edition of *Genetic Variants and Strains of the Laboratory Mouse* (Lyon and Searle, in press). Those rules are regularly updated, and updates are published in *Mouse Genome* (formerly called *Mouse News Letter*; Oxford University Press) and are available on-line in MGD, the Mouse Genome Database. Information on MGD can be obtained from the Mouse Genome Informatics Group, The Jackson Laboratory, Bar Harbor, ME 04609 (telephone, 207-288-3371; fax, 207-288-5079; Internet, mgi-help@informatics.jax.org). The rules for rats have been published as an appendix to the report *Definition, Nomenclature, and Conservation of Rat Strains* (NRC, 1992a), and updates will be published in *Rat Genome*, Heinz W. Kunz, Ph.D., editor, Department of Pathology, University of Pittsburgh School of Medicine, Pittsburgh, PA 15261. Investigators using other laboratory rodents should follow the rules for mice or rats.

Inbred Strains

An inbred strain is designated by capital letters (e.g., mouse strains AKR and CBA and rat strains BN and LEW). The mouse rules, but not the rat rules, allow the use of a combination of letters and numbers, beginning with a letter (e.g., C3H), although this type of symbol is considered less desirable. Brief symbols (generally one to four letters) are preferred. Exceptions are allowed for strains that are already widely known by designations that do not conform (e.g., mouse strains 101 and 129 and rat strains F344 and DONRYU).

Substrains

An established strain is considered to have divided into substrains when genetic differences are known or suspected to have become established in separate branches. These differences can arise either from residual heterozygosity at the time of branching or from new mutations. A substrain is designated by the full strain designation of the parent strain followed by a slanted line (slash) and an appropriate substrain symbol, as follows:

• *Mice*. The substrain symbol can be a number (e.g., DBA/1 and DBA/2); a laboratory code, which is defined below (e.g., C3H/He, where He is the laboratory code for Walter E. Heston); or, when one investigator or laboratory originates more than one substrain, a combination of a number and a laboratory code, beginning with a number (e.g., C57BL/6J and C57BL/10J, where J is the laboratory code for the Jackson Laboratory, Bar Harbor, Maine). Exceptions, such as lower-case letters, are allowed for already well-known substrains (e.g., BALB/c and C57BR/cd).

• *Rats*. The substrain symbol is always a number when genetic differences have been demonstrated. The founding strain is considered the first substrain, and the use of /1 for it is optional (e.g., KGH or KGH/1). A laboratory code (e.g., Pit for the University of Pittsburgh Department of Pathology and N for the NIH Genetic Resource) is used to designate a substrain when genetic differences are probable but not demonstrated (e.g., BN/Pit and BN/N).

Laboratory Codes

Each laboratory or institution that breeds rodents should have a laboratory code. The registry of laboratory codes is maintained by ILAR, National Research Council, 2101 Constitution Avenue, Washington, DC 20418 (telephone, 202-334-2590; fax, 202-334-1687; URL:http://www2.nas.edu/ilarhome/). The laboratory code, which can be used for all laboratory rodents, consists of either a single roman capital letter or an initial roman capital letter and one to three lower-case letters.

• *Mice*. A particular colony is indicated by appending an "@" sign and the laboratory code to the end of the strain or substrain symbol (e.g., SJL@J, the colony of strain SJL mice bred at the Jackson Laboratory; C3H/He@N, the He substrain of strain C3H bred at the NIH Genetic Resource; and CBA/Ca-*se*@J, the Ca substrain of strain CBA carrying the *se* mutation and bred at the Jackson Laboratory). If the substrain symbol and laboratory code are the same, the @ symbol and the laboratory code can be dropped for simplicity (e.g., SJL/J@J becomes SJL/J). The laboratory code is always the last symbol used and is meant to indicate that the environmental conditions and previous history of a colony are unique. When a strain is transferred to a new laboratory, the laboratory code of the originating laboratory is dropped, and the code of the recipient is appended; laboratory codes are not accumulated.

• *Rats*. Normally, a rat strain is designated by the strain name, a slash, the substrain designation (if any), and the laboratory code (e.g., BN/1Pit). When a strain is established in another laboratory, the new laboratory code is appended (e.g., BN/1PitN). In general, more than two laboratory codes are not accumulated. Intermediate codes are dropped to avoid excessively long designations.

For both mice and rats, a strain's holder is responsible for maintaining a strain history.

F1 Hybrids

An F1 hybrid is designated by the full strain designation of the female parent, a multiplication sign, the full strain designation of the male parent, and F1 (e.g., the hybrid mouse C57BL/6J × DBA/2J F1 and the hybrid rat F344/NNia × BN/RijNia F1). If there is any chance of confusion, parentheses should be used to enclose the parental strain names [e.g., (C57BL/6J × DBA/2J)F1 and (F344/NNia × BN/RijNia)F1]. The correct formal name should be given the first time the hybrid is mentioned in a publication; an abbreviated name can be used subsequently [e.g., C57BL/6J × DBA/2J F1 (hereafter called B6D2F1) and F344/NNia × BN/RijNia F1 (hereafter called FBNF1)].

Coisogenic, Congenic, and Segregating Inbred Strains

In mice, a coisogenic strain is designated by the strain symbol, the substrain symbol (if any), a hyphen, and the gene symbol in italics (e.g., CBA/H-*kd*). When the mutant or introduced gene is maintained in the heterozygous condition, this is indicated by including a slash and a plus sign in the symbol (e.g., CBA/H-*kd*/+). A congenic strain is designated by the full or abbreviated symbol of the background strain, a period, an abbreviated symbol of the donor strain, a hyphen, and the symbol of the differential locus and allele (e.g., B10.129-$H12^{b}$). Segregating inbred strains are designated like coisogenic strains; however, indication of the segregating locus is optional when it is part of the standard genotype of the strain (e.g., 129/J and 129/J-c^{ch}/c mean the same thing, and either can be used).

In rats, a coisogenic strain (except for alloantigenic systems—see NRC, 1992a) is designated like a coisogenic strain in mice, except that the laboratory code follows the substrain symbol and the gene symbol is not italicized (e.g., RCS/SidN-rdy). A congenic rat strain (except for alloantigenic systems) is designated like a coisogenic strain (e.g., LEW/N-rnu). For segre gating inbred strains developed by inbreeding with forced heterozygosis, indication of the segregating locus is optional.

Recombinant Inbred (RI) Strains

The symbol of an RI strain should consist of an abbreviation of both parental-strain symbols separated by a capital X with no intervening spaces (e.g., CXB for an RI strain developed from a cross of BALB/c and C57BL mouse strains and LXB for an RI strain developed from a cross of LEW and BN rat strains). Different RI strains in a series should be distinguished by numbers (e.g., CXB1 and CXB2 in mice and LXB1 and LXB2 in rats).

Genes

The rules for gene nomenclature are very complicated because they apply not only to mutant genes, but also to gene complexes, biochemical variants, and other special classes of genes (e.g., transgenes). This description will cover only a small portion of the gene nomenclature. The full rules can be found in the references given previously.

The symbols for loci are brief and are chosen to convey as accurately as possible the characteristic by which the gene is usually recognized (e.g., coat color, a morphologic effect, a change in an enzyme or other protein, or resemblance to a human disease). Symbols for loci are typically two- to four-letter abbreviations of the name. For mice, the symbols are written in italics; for rats, they are not. For convenience in alphabetical lists, the initial letter of the name is usually the same as the initial letter of the symbol. Arabic numbers are included for proteins in which a number is part of the recognized name or abbreviation (e.g., in mice, *C4* and *C6*, the fourth and sixth components of complement, respectively; in rats, C4 and C6). Except in the case of loci discovered because of a recessive mutation, the initial letter of the locus symbol is capitalized and all other letters are lower-case. Hyphens are used in gene symbols only to separate characters that together might be confusing. This rule was adopted for mice in 1993, and hyphens should be deleted from all gene symbols except where they are necessary to avoid confusion. Gene designations are appended to the designation of the parental strain, and they are separated by a hyphen.

Loci That Are Members of a Series

A locus that is a member of a series whose members specify similar proteins or other characteristics is designated by the same letter symbol and a distinguishing number (e.g., *Es1*, *Es2*, and *Es3* in mice and Es1, Es2, and Es3 in rats). For morphologic or "visible" loci with similar effects (e.g., genes that cause hairlessness), distinctive names are given because the gene actions and gene products can ultimately prove to be different (e.g., *hr* and *nu* in mice and fz and rnu in rats).

Alleles

An allele is designated by the locus symbol with an added superscript. For mice, the superscript is written in italics; for rats, it is not. An allele superscript is typically one or two lower-case letters that convey additional information about the allele. For mutant genes, no superscript is used for the first discovered allele. When further alleles are found, the first is still designated without a superscript (e.g., *nu* for nude and nu^{str} for streaker in mice and fa for fatty and fa^{cp} for corpulent in rats). If the information is too complex to be conveyed conveniently in the symbol, the allele is given a superscript (e.g., $Es1^{a}$ and $Es1^{b}$ in mice and $Es1^{\mathrm{a}}$ and $Es1^{\mathrm{b}}$ in rats), and the information is otherwise conveyed. Indistinguishable alleles of independent origin (e.g., recurrences) are designated by the gene symbol with a series symbol, consisting of an Arabic number corresponding to the serial number of the recurring allele plus the laboratory code, appended as a superscript in italics. To avoid confusing the number "1" and the lower-case letter "l," the first discovered allele is left unnumbered, and the second recurring allele is numbered 2 (e.g., *bg*, beige; bg^{J}, a recurrence of the mouse mutation *bg* at the Jackson Laboratory; and bg^{2J}, a second recurrence of the mutation *bg* at the Jackson Laboratory).

A mutation or other variation that occurs in a known allele (except for alloantigenic systems in the rat) is designated by a superscript *m* and an appropriate series symbol, which consists of a number corresponding to the serial number of the mutant allele in the laboratory of origin plus the laboratory code. The symbol is separated from the original allele symbol by a hyphen (e.g., $Mup1^{a\text{-}m1J}$ for the first mutant allele of mouse $Mup1^{a}$ found by the Jackson Laboratory). For a known deletion of all or part of an allele, the superscript *m* may be replaced with the superscript *dl*. This nomenclature is used for naming targeted mutations (often called "knockout" mutations), as well as spontaneously occurring ones.

Transgenes

Nomenclature for transgenes was developed by the ILAR Committee on Transgenic Nomenclature (NRC, 1992b). A transgene symbol consists of three parts, all in roman type, as follows:

TgX(YYYYYY)#####Zzz,

where TgX is the mode, (YYYYYY) is the insert designation, and #####Zzz represents the laboratory-assigned number (#####) and laboratory code (Zzz).

The mode designates the transgene and always consists of the letters Tg (for "transgene") and a letter designating the mode of insertion of the DNA: N for nonhomologous recombination, R for insertion via infection with a retroviral vector, and H for homologous recombination. The purpose of this designation is to identify it as a symbol for a transgene and to distinguish between the three fundamentally different organizations of the introduced sequence relative to the host genome. When a targeted mutation introduced by homologous recombination does not involve the insertion of a novel functional sequence, the new mutant allele (the knockout mutation) is designated in accordance with the guidelines for gene nomenclature for each species. The gene nomenclature is also used when the process of homologous recombination results in integration of a novel functional sequence, if that sequence is a functional drug-resistance gene. For example, Mbp^{m1Dn} would be used to denote the first targeted mutation of the myelin basic protein (*Mbp*) in the mouse made by Muriel T. Davisson (Dn). In this example, the transgenic insertion, even if it contains a functional neomycin-resistance gene, is incidental to "knocking out" or mutating the targeted locus (see also International Committee on Standardized Genetic Nomenclature for Mice, 1994b).

The insert designation is a symbol for the salient features of the transgene, as determined by the investigator. It is always in parentheses and consists of no more than eight characters: letters (capitals or capitals and lower-case letters) or a combination of letters and numbers. Italics, superscripts, subscripts, internal spaces, and punctuation should not be used. Short symbols (six or fewer characters) are preferred. The total number of characters in the insert designation plus the laboratory-assigned number may not exceed 11 (see below); therefore, if seven or eight characters are used, the number of digits in the laboratory-assigned number will be limited to four or three, respectively.

The third part of the symbol is a number and letter combination that uniquely identifies each independently inserted sequence. It is formed of two components. The laboratory-assigned number is a unique number that is assigned by the laboratory to each stably transmitted insertion when germline transmission is confirmed. As many as five characters (numbers as high as 99,999) may be used; however, the total number of characters in the insert designation plus the laboratory-assigned number may not exceed 11. No two lines generated within one laboratory should have the same assigned number. Unique numbers should be given even to separate lines with the same insert integrated at different positions. The number can have some intralaboratory meaning or simply be a number in a series of transgenes produced by the laboratory. The second component is the laboratory code. Thus, the complete designation identifies the inserted site, provides a symbol for ease of communication, and supplies a unique identifier to distinguish it from all other insertions [e.g., C57BL/6J-TgN(CD8Ge)23Jwg for the human CD8 genomic clone inserted into C57BL/6 mice from the Jackson Laboratory (J) and the 23rd mouse screened in a series of microinjections done in the laboratory of Jon W. Gordon (Jwg)]. The complete rules for naming transgenes have been published (NRC, 1992b).

TBASE, a database developed at Oak Ridge National Laboratory, Oak Ridge, Tennessee, as a registry of transgenic strains, is maintained at the Johns Hopkins University, Baltimore, Maryland. Information on TBASE can be obtained from the Genome Database and Applied Research Laboratory, The Johns Hopkins University, 2024 East Monument Street, Baltimore, MD 21205 (telephone, 410-955-1704; fax, 410-614-0434).

Outbred Stocks

An outbred-stock designation consists of a laboratory code, a colon, and a stock symbol that consists of two to four capital letters (e.g., mouse stock Crl:ICR and rat stock Hsd:LE). The stock symbol must not be the same as that for an inbred strain of the same species. As an exception, a stock derived by outbreeding a formerly inbred strain may continue to use the original symbol; in this case, the laboratory code preceding the stock symbol characterizes the stock as outbred. An outbred stock that contains a specified mutation is designated by the laboratory code, a colon, the stock symbol, a hyphen, and the gene symbol (e.g., Crl:ZUC-fa).

The transfer of an outbred stock between breeders is indicated by listing the laboratory code of the new holder followed by the laboratory code of the holder the stock was obtained from (e.g., HsdBlu:LE for rats obtained by Harlan Sprague Dawley from Blue Spruce Farms). To avoid excessively long designations, only two laboratory codes should be used.

QUALITY

In selecting rodents for use in biomedical research, consideration should be given to the quality of the animals. Quality is most commonly characterized in terms of microbiologic status and of the systems used in raising animals to ensure that a specific microbiologic status is maintained. However, the genetics of an animal, as well as the genetic monitoring and breeding programs used to ensure genetic consistency, clearly also play an important part in defining rodent quality.

Microbiologic Quality

Rodents can be infected with a variety of adventitious pathogenic and opportunistic organisms that under the appropriate circumstances can influence research results at either the cellular or subcellular level. Some of those agents can persist in animals throughout their lives; others cause transient infections and are eliminated from the animals, leaving lasting serologic titers as the only indicators that the organisms were present. The types of organisms that can infect rodents include bacteria, protozoa, yeasts, fungi, viruses, rickettsia, mycoplasma, and such nonmicrobial agents as helminths and arthropods.

Many of the common organisms that infect laboratory rodents have been studied extensively, and some of their research interactions have been characterized (see Bhatt et al., 1986;NRC, 1991, for review). Unfortunately, information about the effects of many other organisms is incomplete or is not available. There is no general agreement on the importance of many organisms that latently infect rodents, especially opportunistic organisms that cause disease or alter research results only under narrowly defined conditions and even then usually affect only a very small proportion of the population. Any decision on the quality of rodents to be selected for a particular research project should include a realistic assessment of the organisms that have a reasonable probability, as determined by documentation in the peer-reviewed literature, of producing confounding effects in the proposed study.

It is commonly assumed that animals for which the most extensive health monitoring has been done and to which the most rigorous techniques for excluding microorganisms have been applied are the most appropriate for use in all studies. However, for both scientific and practical reasons, that assumption is not always valid. Rodents that are free of all microorganisms (axenic rodents, see definition below) or axenic rodents that have purposely been inoculated with a few kinds of nonpathogenic microorganisms (microbiologically associated rodents) can have altered physiologic and metabolic processes that make them inappropriate models for some studies. They can also rapidly become contaminated with common microorganisms unless they are maintained with specialized housing and husbandry measures, which are expensive and can fail. The commercial availability of such rodents is limited, and they are more expensive than rodents in which the microbial burden is not so restricted. For those reasons, the rodents most commonly used in research are ones that are free of a few specific rodent pathogens and some other microorganisms that are well known to have confounding effects on specific kinds of research.

The quality of laboratory animals is generally related to the microbiologic exclusion methods used to breed and maintain them. There are three major types of maintenance: isolator-maintained, barrier-maintained, and no-containment or conventionally maintained animals. An isolator is a sterilizable chamber that is usually constructed of metal, rigid plastic, vinyl, or polyurethane. It usually has a sterilized air supply, a mechanism for introducing sterilized materials, and a series of built-in gloves to allow manipulation of the animals housed within. All materials moved into the isolator are sterilized, and animals raised within the isolator are generally maintained free from contamination by either all or specified microorganisms.

Barrier-maintained animals are bred and kept in a dedicated space, called a barrier. For barrier facilities, personnel enter through a series of locks and are usually required to disrobe, shower, and use clean, disinfected clothing. All body surfaces that will potentially make contact with animals are covered. All equipment, supplies, and conditioned air provided to the barrier facility are sterilized or disinfected. Barrier facilities can be of any size and can consist of one or more rooms. They are designed to exclude organisms for which rodents are the primary or preferred hosts but generally will not exclude organisms for which humans are hosts.

Barrier maintenance can also be achieved at the cage or rack level with equipment that can be sterilized or otherwise disinfected. This type of maintenance depends heavily on providing large volumes of filtered or sterilized air to the animal cages. Such systems can be used successfully to maintain animals with a highly defined microbiologic status; the success of such systems depends on the techniques used and is difficult to monitor because microbiologic status might differ from cage to cage.

No-containment, or conventionally maintained, animals are raised in areas that have no special impediments to the introduction of microorganisms. This method of maintaining animals cannot ensure stability of the microbiologic status, because unwanted organisms can be introduced at any time.

Several classifications have been developed to define the microbiologic quality of laboratory animals, as follows (see also NRC, 1991):

- *Axenic* refers to animals that are derived by cesarean section or embryo transfer and reared and maintained in an isolator with aseptic techniques. It implies that the animals are demonstrably free of associated forms of life, including viruses, bacteria, fungi, protozoa, and other saprophytic or parasitic organisms. Animals of this quality require the most comprehensive and frequent monitoring of their microbiologic status and are the most difficult to obtain and maintain.
- *Microbiologically associated, defined flora,* or *gnotobiotic* refers to axenic animals that have been intentionally inoculated with a well-defined mixture of microorganisms and maintained continuously in an isolator to prevent contamination by other agents. Generally, a small number (usually less than 15) of species of microorganisms are used in the inoculum, and it is implied that these organisms are nonpathogenic.
- *Pathogen-free* implies that the animals are free of all demonstrable pathogens. It is often misused, in that there is no general agreement about which agents are pathogens, what tests should be used to demonstrate the lack of pathogens and with what frequency, and how the populations should be sampled. Use of this term should be avoided because of the lack of precision of its meaning.
- *Specific-pathogen-free* (SPF) is applied to animals that show no evidence (usually by serology, culture, or histopathology) of the presence of particular microorganisms. In its strictest sense, the term should be related to a specific set of organisms and a specific set of tests or methods used to detect them. An animal can be classified as SPF if it is free of one or many pathogens.
- *Conventional* is applied to animals in which the microbial burden is unknown, uncontrolled, or both.

In addition, the term *clean conventional* is sometimes used to describe animals that are maintained in a low-security barrier and are demonstrated to be free of selected pathogens. This term is even less precise than *pathogen-free*, and its use is discouraged (NRC, 1991).

Commercial suppliers have coined various terms to indicate SPF status. All the terms are related to specific organisms of which the animals are stated to be free and for which they are regularly monitored. In some cases, the terms (e.g., *virus-antibody-free* and *murine-pathogen-free*) imply a quality of animals beyond the actual definitions of the terms. Virus-antibody-free animals, for example, are animals that are free of antibodies to specific rodent viruses. The term is a variation of SPF, in that it relates to *specified* viruses. The implied method of detection is serology. Animals might not be free of viruses other than those specified and might not be free of other microorganisms.

Genetic Quality

In spite of diligent maintenance practices that are required in any breeding colony to identify animals properly and house them securely, people can make mistakes. In addition, loose animals, including animals that escape their housing unnoticed and wild rodents, can enter cages, mate with the inhabitants, and produce genetically contaminated offspring. Good husbandry practices carried out by trained personnel, including keeping a pedigree and clearly identifying animals and cages, can help to reduce the occurrence of such events. Nevertheless, to avoid devastating consequences of genetic contamination, a good program of genetic monitoring is warranted. Genetic monitoring consists of any method used to ensure that the genetic integrity of individuals of any particular strain has not been violated. Several commercial sources provide genetic monitoring services for inbred mouse and rat strains.

Personnel should be alert to phenotypic changes in the animals, such as unexpected coat colors or large changes in reproductive performance. In a pedigree-controlled foundation colony (see Chapter 4), it is important to monitor the breeding stock at least once every two generations so that a single erroneous mating can be detected quickly. Retired breeders or some of their progeny can be tested. In an expansion or production colony, in which it might not be cost-effective or practical to monitor so closely, sampling is recommended. The extent of such sampling can be as broad as resources and need permit. If genetic contamination occurs outside the foundation colony, contamination will eventually be purged by the infusion of breeders from the more rigorously controlled foundation colony.

The extent of necessary testing depends on the number and genotypes of neighboring strains. A testing system should be capable of identifying the strain to which the individual belongs and differentiating it from other strains maintained nearby. Most strains can be identified with a small set of any genetic markers for which an assay is available. Newer DNA-typing methods that use multilocus probes, minisatellite markers, and "DNA-fingerprinting" analysis are powerful tools for distinguishing strains, especially strains that are closely related, but electrophoretic methods that type isoenzymes are generally more cost-effective for genetic monitoring (Hedrich, 1990; Nomura et al., 1984), in that such monitoring is most commonly done to detect mismatings. Immunologic methods are also used, and the exchange of skin grafts between individuals of a strain is a particularly effective method for screening a large number of loci in a single test. DNA from representative breeders of a strain can be stored for future use in identifying suspected genetic contaminations.

Genetic monitoring is used primarily to verify the authenticity of a given strain; new mutations are rarely detected by this means. It is impossible to monitor all loci for new mutations, given the large number of unknown loci and known loci that do not produce a visible phenotype. A good breeding-management program, as described in Chapter 4, will help to reduce unwanted genetic changes caused by mutations.

SELECTED ASPECTS OF EXPERIMENTAL DESIGN

An experiment in which laboratory animals are used should be designed carefully, so that it produces unequivocal information about the questions that it was designed to address. The two most important requirements of proper experimental design in that connection are as follows:

- Animals in different groups should vary only in the treatment that the experiment is designed to evaluate, so that the experimental outcome will not be confounded by dissimilarities in the constitution of the groups or in how they are treated or measured.
- Each treatment should be given to enough animals for the experimental outcome to be attributed confidently to treatment difference and not merely to chance.

The best way to ensure that groups of experimental animals are comparable is to draw them from a single homogeneous pool and to assign them randomly to treatment groups. Choosing animals of the same age, sex, and inbred strain for all treatment groups and even assigning littermates randomly to different treatment groups can eliminate factors that might partially account for group-to-group differences in experimental outcome.

Once animals are assigned to groups, they should be handled identically, except for the treatment differences that the experiment is designed to evaluate. Food, water, bedding, and other features of animal husbandry should be the same. For long-term experiments, cages should be rotated to minimize group differences caused by cage position. For invasive experimental treatments, sham or placebo procedures should be performed in comparison groups; for example, animals given treatment by gavage should be compared with controls given the vehicle by gavage, animals treated surgically should be compared with animals that undergo sham surgical operations, and animals exposed to treatment by inhalation should be compared with animals placed in inhalation chambers that circulate only air. Following those precautions will ensure that differences in outcome between groups can be attributed to the experimental treatment itself and not to ancillary differences associated with the administration of the treatment.

Finally, wherever possible, the outcome of interest should be measured by people who are unaware of which treatment each animal received, because such knowledge can magnify or even create observed treatment differences. It is particularly important to carry out "blind" studies when the outcome is to be evaluated subjectively (e.g., by grading of disease severity), rather than measured quantitatively (e.g., by measuring concentrations of serum constituents).

The number of animals needed in each group will depend on many features of the experimental design, including the following:

- the goals of the study;
- the primary outcome measure that will be compared;
- the number of groups that will be compared;
- the expected number of technical failures or usable end points;
- the number and type of comparisons that will be made;
- the expected animal-to-animal and measurement variability in the outcome;
- the statistical design and analysis that will be used;
- the magnitude of the differences between control and treatment groups that it is desirable to detect;
- the projected losses; and
- the maximal tolerable chance of drawing erroneous conclusions.

The more variable an outcome measure is, either because outcomes in identically treated animals vary substantially or because there is a high degree of measurement variability, the more animals will be needed in each group to distinguish between group differences caused by treatment and those caused by chance. How outcome measurement variability, treatment difference to be detected, and tolerable chance of drawing an erroneous conclusion affect the required sample size depends on the measurement to be made, the type of group comparison to be made, and the statistical analysis to be used. Tables and formulas for comparing proportions among two or more groups have been published (Gart et al., 1986), as has useful information for other types of outcomes (Mann et al., 1991). For most experiments, it is highly desirable to collaborate with a statistician throughout, beginning with the design stage, so that appropriately defined groups of sufficient size will be available for a proper statistical analysis.

REFERENCES

Altman, P. L., and D. D. Katz, eds. 1979a. Inbred and Genetically Defined Strains of Laboratory Animals. Part I: Mouse and Rat. Bethesda, Md.: Federation of American Societies for Experimental Biology. 418 pp.

Altman, P. L., and D. D. Katz, eds. 1979b. Inbred and Genetically Defined Strains of Laboratory Animals. Part II: Hamster, Guinea Pig, Rabbit, and Chicken. Bethesda, Md.: Federation of American Societies for Experimental Biology. 319 pp.

Bailey, D. W. 1981. Recombinant inbred strains and bilineal congenic strains. Pp. 223-239 in The Mouse in Biomedical Research. Vol. I: History, Genetics, and Wild Mice, H. L. Foster, J. D. Small, and J. G. Fox, eds. New York: Academic Press.

Baker, H. J., J. Russell Lindsey, and S. H. Wiesbroth, eds. 1979-1980. The Laboratory Rat. Vol. I, Biology and Diseases, 1979, 435 pp.; Vol. II, Research Applications, 1980, 276 pp. New York: Academic Press.

Bhatt, P. N., R. O. Jacoby, H. C. Morse III, and A. E. New, eds. 1986. Viral and Mycoplasmal Infections of Laboratory Rodents: Effects on Biomedical Research. Orlando, Fla.: Academic Press.

Demant, P., A.A. Hart. 1986. Recombinant congenic strains—A new tool for analyzing genetic traits determined by more than one gene. Immunogenetics 24(6):416-422.

Festing, M. F. W. 1989. Inbred strains of mice. Pp. 636-648 in Genetic Variants and Strains of the Laboratory Mouse, 2d ed, M. F. Lyon and A. G. Searle, eds. Oxford: Oxford University Press.

Festing, M. F. W. 1993. International Index of Laboratory Animals, 6th ed. Leicester, U.K. M. F. W. Festing. 238 pp. Available from M. F. W. Festing, PO Box 301, Leicester LE1 7RE, UK.

Festing, M. F. W., and D. D. Greenhouse. 1992. Abbreviated list of inbred strains of rats. Rat News Letter 26:10-22.

Foster, H. L., J. D. Small, and J. G. Fox, eds. 1981-1983. The Mouse in Biomedical Research. Vol. I: History, Genetics, and Wild Mice, 1981, 306 pp.; Vol. II: Diseases, 1982, 449 pp.; Vol. III: Normative Biology, Immunology, and Husbandry, 1983, 447 pp.; Vol. IV: Experimental Biology and Oncology, 1982, 561 pp. New York: Academic Press.

Fox, J. G., B. J. Cohen, and F. M. Lowe, eds. 1984. Laboratory Animal Medicine. Orlando, Fla.: Academic Press. 750 pp.

Gart, J. J., D. Krewski, P. N. Lee, R. E. Tarone, and J. Wahrendorf. 1986. Statistical methods in cancer research. Volume III: The design and analysis of long-term animal experiments. Pub. No. 79. IARC Scientific Publications.

Gill, T. J. 1980. The use of randomly bred and genetically defined animals in biomedical research. Am. J. Pathol. 101(3S):S21-S32.

Gill, T. J., III, G. J. Smith, R. W. Wissler, and H. W. Kunz. 1989. The rat as an experimental animal. Science 245:269-276.

Green, E. L. 1981. Genetics and Probability in Animal Breeding Experiments. New York: Oxford University Press. 271 pp.

Hansen, C. T., S. Potkay, W. T. Watson, and R. A. Whitney, Jr. 1981. NIH Rodents: 1980 Catalogue. NIH Pub. No. 81-606. Washington, D.C.: U.S. Department of Health and Human Services. 253 pp.

Harkness, J. E., and J. E. Wagner. 1989. The Biology and Medicine of Rabbits and Rodents, 3rd ed. Philadelphia: Lea & Febiger. 230 pp.

Hedrich, H. J., M. Adams, ed. 1990. Genetic Monitoring of Inbred Strains of Rats: A Manual on Colony Management, Basic Monitoring Techniques, and Genetic Variants of the Laboratory Rat. Stuttgart: Gustav Fischer Verlag. 539 pp.

Hogan, B., F. Costantini, and E. Lacy. 1986. Manipulating the Mouse Embryo: A Laboratory Manual. Cold Spring Harbor, N.Y.: Cold Spring Harbor Laboratory. 332 pp.

International Committee on Standardized Genetic Nomenclature for Mice. 1994a. Rules for nomenclature of inbred strains. Mouse Genome 92(2):xxviii-xxxii.

International Committee on Standardized Genetic Nomenclature for Mice. 1994b. Rules and guidelines for gene nomenclature. Mouse Genome 92(2):viii-xxiii.

Mann, M. D., D. A. Crouse, and E. D. Prentice. 1991. Appropriate animal numbers in biomedical research in light of animal welfare considerations. Lab. Animal Sci. 41(1):6-14.

Merlino, G. T. 1991. Transgenic animals in biomedical research. FASEB J. 5:2996-3001.

Nomura, T., K. Esaki, and T. Tomita, eds. 1984. ICLAS Manual for Genetic Monitoring of Inbred Mice. Tokyo: University of Tokyo Press.

NRC (National Research Council), Institute of Laboratory Animal Resources, Committee on Immunologically Compromised Rodents. 1989. Immunodeficient Rodents: A Guide to Their Immunobiology, Husbandry, and Use. Washington, D.C.: National Academy Press. 246 pp.

NRC (National Research Council), Institute of Laboratory Animal Resources, Committee on Infectious Diseases of Mice and Rats. 1991. Infectious Diseases of Mice and Rats. Washington, D.C.: National Academy Press. 397 pp.

NRC (National Research Council), Institute of Laboratory Animal Resources, Committee on Rat Nomenclature. 1992a. Definition, nomenclature, and conservation of rat strains. ILAR News 34(4):S1-S26.

NRC (National Research Council), Institute of Laboratory Animal Resources, Committee on Transgenic Nomenclature. 1992b. Standardized nomenclature for transgenic animals. ILAR News 34(4):45-52.

Van Hoosier, G. L., Jr., and C. W. McPherson, eds. 1987. Laboratory Hamsters. Orlando, Fla.: Academic Press. 400 pp.

Wagner, J. E., and P. J. Manning, eds. 1976. The Biology of the Guinea Pig. New York: Academic Press. 317 pp.

4

Genetic Management of Breeding Colonies

Different breeding systems and genetic-engineering methods have been used to produce strains and stocks of rodents for particular experimental purposes—inbred strains; coisogenic, congenic, and transgenic strains; recombinant inbred strains; hybrid strains; and outbred stocks. Outbred stocks are used primarily when genetic heterogeneity is desired and are not useful when a controlled genotype is required. However, the loss of heterozygosity cannot be completely avoided in propagating outbred stocks, because the breeding population is necessarily finite.

GENETICALLY DEFINED STOCKS

Regardless of the breeding system or genetic manipulation used to produce a particular strain, some practices are recommended to maintain high genetic quality. Details of breeding systems used to develop various types of strains can be found elsewhere (Bailey, 1981; Green, 1981a). Here we describe the management of breeding colonies of already-developed strains.

Pedigrees

Using a pedigree method allows the parentage of individual experimental animals to be traced; aids in selection of parental pairs to avoid the inadvertent fixation of unwanted mutations, especially mutations that would affect reproductive performance; and maximizes genetic uniformity within a strain.

Traceability

Mutations occur continually in any breeding stock. Many of these mutations are recessive and, when homozygous, will be expressed as undesirable traits. When such a mutation is expressed, it is necessary to rid the breeding colony of copies of the mutation that might be carried as a heterozygous gene by individuals that are normal in phenotype. Use of a pedigree system that records the parents of each individual makes it possible to identify relatives of the affected individual, and they can be tested for the presence of the mutation or eliminated from the colony. It is also desirable to mark the animals with their pedigree identification.

Selection of Parental Pairs

Reproductive performance, even within a highly inbred strain, can vary greatly. Environmental factors undoubtedly cause much of that variation, but spontaneously occurring mutations that adversely affect breeding performance are also contributing factors. To avoid extinction of a strain, the individuals selected for propagating it should be those with the best reproductive performance. Reproductive performance can be evaluated retroactively by examining a pedigree, that is, the reproductive performance of several generations of offspring can be used in evaluating the breeding performance of the original pair and can aid in avoiding the accidental incorporation or accumulation of deleterious recessive mutations. To ensure continuation of a strain, several families or lines should be maintained for two to three generations until one pair in each generation is retroactively chosen as the pair from which breeders in all subsequent generations will be derived. This practice not only ensures selection of reproductively fit individuals to propagate the strain but also maximizes genetic uniformity, as described below.

Genetic Uniformity

The purpose of producing an inbred strain is to achieve genetic uniformity among individuals. That allows a greater degree of reproducibility in experiments than is possible if heterogeneous individuals are used. However, total genetic uniformity is never achieved, because new mutations occur. Each new mutation has a 25 percent chance of becoming fixed in an inbred strain (Bailey, 1979). The gradual accumulation of such mutations and the resulting genetic changes are called *genetic drift*. Because of the random occurrence of mutations, genetic drift will involve different genes in two separately maintained sublines of a strain. Over time, the sublines will become increasingly different from each other; this tendency is called *subline divergence*. Bailey has estimated that separately maintained sublines will diverge at the rate of approximately one new mutation every two generations (Bailey, 1978, 1979, 1982). Even within one breeding colony, subline divergence can occur if the propagation of family branches is allowed to continue indefinitely.

Another source of subline differences is the genetic heterogeneity present in a strain at the time of subline separation. Many of the early substrains of common inbred strains were separated before the strain had been highly inbred; for example, mouse substrains C57BL/6 and C57BL/10 were separated from the C57BL strain when it had been inbred for only about 30 generations. That is more than the 20 generations conventionally accepted as the definition of an inbred strain, but the amount of heterogeneity, although small in comparison with the total number of genes, is still sufficient to result in subline differences. For example, according to Bailey's estimates, one would expect about 14 fixed differences between substrains C57BL/6 and C57BL/10 caused by the presence of unfixed genes at the time of separation. Bailey also showed that the probability of there being no heterogeneity within an inbred strain does not reach 0.99 until after 60 generations of brother × sister inbreeding (Bailey, 1978). The practical consequence of subline divergence for research is that animals from different sublines might respond differently in identical experiments, and the difference in responses could lead to misinterpretation of the experimental results. A corollary is that no subline (or substrain) can be considered a reference standard, because all sublines undergo changes with time. Cryopreservation might offer the only means to arrest such changes. Nevertheless, it is wise to obtain breeders periodically from the original source colony, to maximize homogeneity between two colonies. A general practice is to do that after 10 generations of separation.

Within a breeding colony, pedigree management can be used to maximize genetic uniformity. One pair in each generation can be selected on the basis of breeding performance, to be the common ancestral mating for all progeny. So that all animals at any time can be traced to a single ancestral pair, the number of generations of any branch other than the common ancestral branch is limited, depending on the number of animals that are produced for experimental use, the productivity or the average number of breeding pairs of progeny expected from a single mating, and the reproductive life span of breeders.

Because most commonly used inbred strains today are highly inbred, breeding selection is not effective in increasing reproductive performance. Rather, selection is made to avoid deleterious mutations that would cause a decrease in reproductive performance. The prevalence and rate of such mutations are unknown, but distinct reductions in reproductive performance within family branches have been observed in large breeding colonies. Because increases in reproductive performance are rare, mutations that are advantageous to reproduction are probably extremely rare.

Pedigree identification of animals used as parents for the production of hybrids is advised so that mutations or irregularities can be traced. However, pedigree management is not necessary, because there is no propagation of lines beyond that of the F1 generation.

Foundation or Nucleus Colonies, Expansion Colonies, and Production Colonies

In large breeding operations, it is often practical for management purposes to subdivide the breeding colony of each strain into separate groups—a foundation colony (sometimes called a nucleus colony), an expansion colony, and a production colony—that are maintained in separate facilities. A foundation colony is a breeding colony of sufficient size to propagate the strain (following the selection procedures described previously) and to provide breeding stock to an expansion colony. The purpose of an expansion colony is to increase the number of breeding pairs to a quantity adequate to support a production colony. A production colony is made up of breeders from an expansion colony; offspring are distributed for research, not used for breeding.

It is more practical to be rigorous about selection practices and genetic monitoring in a foundation colony, which is relatively small, than in the larger expansion and production colonies. It is also more important to carry out those activities in the foundation colony because all the stock in the expansion and production colonies is ultimately derived from it and any change occurring in the foundation colonies will eventually be propagated throughout the entire strain. An advantage of using a separate facility for foundation colonies is that it permits microbiologic status of the foundation colony to be maintained with fewer pathogens than the other colonies. Often, foundation colonies are maintained in a separate building from expansion and production colonies to protect against loss of a strain due to disease outbreak or other catastrophe. Cryopreservation and storage of embryos can also fulfill that security requirement.

In an expansion colony, it might not be practical or cost-effective to maintain detailed pedigree records or devote much time to selection. It is relatively easy, however, to keep track of the number of generations that a family or subline has been separated from the foundation stock by making a notation on the cage card each time a new mating group is made up. By limiting the number of generations outside the foundation nucleus, maximal genetic uniformity can be achieved. Unnoticed mutations (e.g., those affecting reproductive performance) that occur in either an expansion or a production colony will ultimately be purged because of the constant infusion of highly scrutinized breeding stock from the foundation colony. Trio matings (i.e., two females mated to a sibling male) are often used in expansion colonies for efficiency.

In a production colony, especially a large one, the use of non-sib matings increases efficiency. The probability that recessive, mutated alleles will come together and be expressed in an individual is much decreased when non-sib matings are used. However, it is also less likely that such mutations will be detected and eliminated; therefore, it is not recommended that strains be propagated for more than a few generations by non-sib matings. Normally, breeders in a production colony represent the last generation of family lines created in enlarging the colony.

NONGENETICALLY DEFINED STOCKS

The goal of breeding programs for nongenetically defined stocks is to maintain the diversity in genotypes that is present in the founding animals of that stock. Ideally, no selection pressures should be placed on the population; however, in practice, there is often a conscious or unconscious selection for reproductive performance, and great care should be taken to eliminate this bias. Ideally, a purely random mating structure should be used so that each animal has an equal chance of participating in the breeding program and of mating with any of the animals of the opposite sex within the colony with no attention to relationship, genotype, phenotype, or any other characteristic; this requires accurate identification of individual animals, extensive record-keeping, and structured randomization in which randomization tables or computer-generated randomized numbers are used to select breeding pairs.

An important limitation on any random breeding program is the size of the population that can be maintained within a facility. Even for commercial breeders, populations are limited in size; therefore, without a systematic method for ensuring that inbreeding does not occur, chance matings between relatives will gradually cause a decrease in heterozygosity within the population. The rate of decrease of heterozygosity is proportional to the population size; very small populations experience a more rapid decrease. For example, a population of 50 will undergo a decrease in heterozygosity at the rate of about 1 percent per generation. After 20 generations, this population will have only 82 percent of the heterozygosity with which it started (Green, 1981b).

To minimize that loss of heterozygosity, one can use a structured system of mating that is not completely random but is designed to avoid inbreeding. Several such systems exist. In very small populations (up to 32 animals), systematic mating of cousins can be used to avoid brother × sister mating. When the number of animals exceeds 32, that system becomes too cumbersome to use. In larger colonies, either a circular or circular-paired mating system can be used effectively to minimize inbreeding; both systems slow the loss of heterozygosity and require regular pairing of progeny from individual cages or groups of cages with animals in adjacent cages or groups. Detailed descriptions of these systems are available (Kimura and Crow, 1963; Poiley, 1960). Alternatively, a computerized system of tracking the coefficient of inbreeding of all breeders can be used to set up matings of the least-related animals.

Loss of heterozygosity by inadvertent inbreeding and acquisition and fixation of spontaneous mutations can cause considerable genetic divergence between populations of the same nongenetically defined stock maintained at different locations. To minimize the process, there should be a regular exchange of breeding stock between populations. The number of animals that are transferred and the frequency of transfer will depend on many factors, including colony size, breeding system used, and rate at which divergence is anticipated to occur. The success of such measures can be assessed with population-genetics techniques to calculate the degree of residual heterozygosity in individual populations. These methods usually entail surveying a large number of biochemical or immunologic markers that display polymorphism in a relatively large sample of the population.

In addition to the classic nongenetically defined populations maintained by random breeding or outbreeding, populations of rodents with substantial genetic diversity, as evidenced by heterozygosity at a large number of loci, can be developed by making systematic multiple inbred-strain crosses. In such a system, four or more inbred strains are regularly crossed in a circular fashion to yield F1 progeny that are systematically mated with a rotational system to provide F2 animals for use in experimental procedures. F2 animals will show greater genetic diversity than most common nongenetically defined stocks that have been maintained for many years as closed colonies (Green, 1981b).

Overall, the maintenance of nongenetically defined stocks is complex if inbreeding is to be minimized. These populations are unique, dynamic, and diverse and require regular characterization unless they are linked by exchange of breeding stock.

CRYOPRESERVATION

Cryopreservation, in the form of freezing of cleavage-stage embryos, offers a means to protect a stock or strain against accidental loss or genetic contamination. It also provides a genetic advantage in retarding genetic changes caused by accumulated mutations and an economic advantage in lowering the costs of strain maintenance. In some circumstances, as when quarantine regulations impede the importation of adult animals, the transportation of frozen embryos, which do not have to be quarantined, is effective. Cryopreservation of embryos has been possible since 1972 (Whittingham et al., 1972; Wilmut, 1972) and has now been successfully carried out for at least 16 mammalian species, including mice and rats (Hedrich and Reetz, 1990; Leibo, 1986; Whittingham, 1975; Whittingham et al., 1972).

Not all stocks warrant cryopreservation. If a strain is preserved with scant information on its characteristics, for example, it is unlikely that it will be of much use in the future. The ILAR Committee on Preservation of Laboratory Resources has recommended the following criteria for identifying valuable laboratory animals: the importance of the disease process or physiologic function, the validity or genetic integrity of the stock, the difficulty of replacing the stock, versatility of the stock, and current use (NRC, 1990).

To obtain embryos of a predetermined stage for freezing, exogenous gonadotropins are administered to induce synchronous ovulation and permit timed matings. Exogenous gonadotropins also often induce superovulation (i.e., the production of more eggs than normal). A combination of pregnant mares' serum, which contains follicle-stimulating hormone, and human chorionic gonadotropin, which contains luteinizing hormone, is commonly used (Gates, 1971). Freezing eight-cell embryos generally produces the most reliable results, at least in the mouse, but other preimplantation embryo stages can also be used.

There are many methods for cryopreserving embryos (Leibo, 1992; Mazur, 1990). Generally, they are in two categories: equilibrium methods and nonequilibrium methods; the distinction depends on the osmotic forces encountered in the presence of cryoprotectant during the freezing process (Mazur, 1990). Equilibrium methods use low concentrations (1.5M) of cryoprotectants and slow, controlled cooling (approximately 0.5°C/min). Nonequilibrium methods generally use a higher concentration of cryoprotectants (about 4-5 M) and fast cooling (more than 200°C/min). The two kinds of methods are equally successful, but nonequilibrium methods have the advantage of not requiring controlled-rate freezers.

In mice, 500 is generally considered a safe number of embryos to store. Mouse embryos show no deterioration with time when stored at –196°C, and their viability is not affected by the equivalent of 2,000 years of exposure to background radiation (Glenister et al., 1984, 1990). Mice have been born from embryos stored for 14 years with no observable differences in rates of birth from recently frozen embryos. An advantage of liquid-nitrogen storage systems is that electricity and motors are not required; only a periodic, and preferably routine, replenishment of liquid nitrogen is necessary. Alarms and automatic filling devices need electricity, but all maintenance and monitoring of liquid-nitrogen storage containers can be carried out manually if necessary.

To recover animals from frozen embryos, the embryos are thawed and transferred to pseudopregnant females, that is, females in which the hormones required to support implantation and pregnancy are induced by mating them to vasectomized or genetically sterile males. The overall rate of live births from frozen mouse embryos of inbred and mutant strains is 20 percent. The rate is usually higher for hybrid and outbred embryos, but there is extreme variability, and the rate from a given attempt can range from 0 to 100 percent.

For security, embryos from one strain would ideally be stored in separate cities; at a minimum they should be stored in two containers. Before a strain is considered safely cryopreserved, it should have been re-established at least once from frozen embryos by recovering live born, raising them to maturity, and breeding them to produce the next generation. To avoid genetic contamination of a strain, genetic monitoring procedures should be used to verify that animals born from frozen embryos have the expected genotype.

RECORD-KEEPING

In maintaining pedigrees, the most critical records are those of parentage. One should be able to identify and trace all relationships through these records. In addition to parental information, which might include individual identification numbers and mating dates, it is useful to record the generation number, birthdate, number born, weaning date, number weaned, and disposition of progeny. The latter information is useful in evaluating the reproductive performance of a colony. A bound, archive-quality pedigree ledger or a secure computer system might be used for recording information. A computer program for colony record-keeping has been described (Silver, 1993). If ledgers are used in a colony that includes many strains, it is useful to maintain a separate book for each strain. Each book should identify the book that preceded it or, if it is the first pedigree record for its colony, the origin of the animals. In colonies that have only a few strains, it might be more practical to maintain one general ledger. In this case, it is important to identify each entry accurately according to its strain, as well as its parental and other information. For pedigree management, it is also useful to maintain a pedigree chart, at least for foundation breeders; this helps to avoid unnecessary proliferation of family branches by allowing visualization of individual animal relationships.

Marking of each animal with its pedigree identification will preserve identity throughout its lifetime (see Chapter 5). That can be useful when animals from different sibships are housed in the same cage. The advantage of recording individual identifications of animals used in research is that retrospective analysis of such characteristics as age and family relationship can sometimes help to explain unexpected results.

REFERENCES

Bailey, D. W. 1978. Sources of subline divergence and their relative importance for sublines of six major inbred strains of mice. Pp. 197-215 in Origins of Inbred Mice, H. C. Morse III, ed. New York: Academic Press.

Bailey, D. W. 1979. Genetic drift: The problem and its possible solution by frozen-embryo storage. Pp. 291-299 in The Freezing of Mammalian Embryos, K. Elliott and J. Whelan, eds. CIBA Foundation Symposium 52 (New Series). Amsterdam: Excerpta Medica.

Bailey, D. W. 1981. Recombinant inbred strains and bilineal congenic strains. Pp. 223-239 in The Mouse in Biomedical Research. Vol. I: History, Genetics, and Wild Mice, H. L. Foster, J. D. Small, and J. G. Fox, eds. New York: Academic Press.

Bailey, D. W. 1982. How pure are inbred strains of mice. Immunol. Today 3(8):210-214.

Gates, A. H. 1971. Maximizing yield and developmental uniformity of eggs. Pp. 64-75 in Methods in Mammalian Embryology, J. C. Daniel, Jr., ed. San Francisco: Freeman.

Glenister, P. H., D. G. Whittingham, et al. 1984. Further studies on the effect of radiation during the storage of frozen 8-cell mouse embryos at -196 degrees C. J. Reprod. Fertil. 70:229-234.

Glenister, P. H., D. G. Whittingham, et al. 1990. Genome cryopreservation—A valuable contribution to mammalian genetic research. Genet. Res. 56:253-258.

Green, E. L. 1981a. Genetics and Probability in Animal Breeding Experiments. New York: Oxford University Press. 271 pp.

Green, E. L. 1981b. Breeding systems. Pp. 91-104 in The Mouse in Biomedical Research. Vol. I.: History, Genetics and Wild Mice, H. L. Foster, J. D. Small, and J. G. Fox, eds. New York: Academic Press.

Hedrich, H. J., and I. C. Reetz. 1990. Cryopreservation of rat embryos. Pp. 274-288 in Genetic Monitoring of Inbred Strains of Rats: A Manual on Colony Management, Basic Monitoring Techniques, and Genetic Variants of the Laboratory Rat, H. J. Hedrich, ed. Stuttgart: Gustav Fischer Verlag.

Kimura, M., and J. F. Crow. 1963. On maximum avoidance of inbreeding. Genet. Res. 4:399-415.

Leibo, S. P. 1986. Cryobiology: Preservation of mammalian embryos. Pp. 251-272 in Genetic Engineering of Animals An Agricultural Perspective. J. W. Evans and A. Hollaender, eds. New York: Plenum Press.

Leibo, S. P. 1992. Techniques for preservation of mammalian germ plasm. Anim. Biotechnol. 3(1):139-153.

Mazur, P. 1990. Equilibrium, quasi-equilibrium, and nonequilibrium freezing of mammalian embryos. Cell Biophys. 17:53-92.

NRC (National Research Council), Institute of Laboratory Animal Resources, Committee on Preservation of Laboratory Animal Resources. 1990. Important laboratory animal resources: Selection criteria and funding mechanisms for their preservation. ILAR News 32(4):A1-A32.

Poiley, S. M. 1960. A systematic method of breeder rotation for non-inbred laboratory animal colonies. Proc. Anim. Care Panel 10:159-166.

Silver, L. M. 1993. Recordkeeping and database analysis of breeding colonies. Pp. 3-15 in Guide to Techniques in Mouse Development, P. M. Wassarman and M. L. DePamphilis, eds. Methods in Enzymology, Volume 225. San Diego: Academic Press.

Whittingham, D. G. 1975. Survival of rat embryos after freezing and thawing. J. Reprod. Fertil. 43:575-578.

Whittingham, D. G., S. P. Leibo, and P. Mazur. 1972. Survival of mouse embryos frozen to -196°C and -269°C. Science 178:411-414.

Wilmut, I. 1972. The effect of cooling rate, warming rate of cryoprotective agent, and stage of development on survival of mouse embryos during freezing and thawing. Life Sci. (II),11:1071-1079.

5

Husbandry

HOUSING

Caging

Caging is one of the primary components of a rodent's environment and can influence the well-being of the animals it houses. Many types of caging are available commercially. Those used to house rodents should have the following features:

- They should accommodate the normal physiologic and behavioral needs of the animals, including maintenance of body temperature, normal movement and postural adjustments, urination and defecation, and, when indicated, reproduction.
- They should facilitate the ability of the animal to remain clean and dry.
- They should allow adequate ventilation.
- They should allow the animals easy access to food and water and permit easy refilling and cleaning of the devices that contain food and water.
- They should provide a secure environment that does not allow animals to become entrapped between opposing surfaces or in ventilation openings.
- They should be free of sharp edges or projections that could cause injury to the animals housed.
- They should be constructed so that the animals can be seen easily without undue disturbance.

- They should have smooth, nonporous surfaces that will withstand regular sanitizing with hot water, detergents, and disinfectants.
- They should be constructed of materials that are not susceptible to corrosion.

In selecting caging, one should pay close attention to the ease and thoroughness with which a cage can be serviced and sanitized. In addition to smooth, impervious surfaces that are free of sharp edges, cages should have minimal corners, ledges, and overlapping surfaces, because these features allow the accumulation of dirt, debris, and moisture. Cages should be constructed of durable materials that can withstand rough handling without chipping or cracking.

Sanitizing procedures, such as autoclaving and exposure to ionizing radiation, can alter the physical characteristics of caging materials over time and can greatly shorten useful life. Rodent cages are most commonly constructed of stainless steel or plastic (polyethylene, polypropylene, or polycarbonate), each of which has advantages and disadvantages. Galvanized metal and aluminum have also been used but are generally less acceptable because of their high potential for corrosion.

Most rodent cages have at least one wire or metal grid surface to furnish ventilation and permit inspection of the animals in the cage. Inspection of animals can be further facilitated by the use of transparent plastic cages. Opaque plastic or metal cages might provide a more desirable environment for some studies or breeding programs; however, adequate inspection of animals will usually require manipulation of each cage.

The bottoms of rodent cages can be either solid or wire. The floors of solid-bottom cages usually are covered with bedding material that absorbs urine and moisture from feces, thereby improving the quality of the cage environment and allowing for easy removal of accumulated wastes. Solid-bottom cages provide excellent support for rodents' feet, minimizing the occurrence of pododermatitis and injuries. Wire-bottom cages are equipped with a wire-mesh grid, the spaces in which are large enough to allow the passage of feces. Generally, there are two to four wires per inch (2.5 cm) in the grid. These cages are normally mounted on racks that suspend them over waste-collection pans filled with absorbent material. This caging type minimizes contact with feces and urine and is thought to improve cage ventilation. However, careful consideration should be given to the size and species of rodents to be housed in wire-bottom cages because if their feet and legs can be entrapped in the wire grid, they can suffer severe trauma, including broken bones. In addition, older, heavier rodents can develop pododermatitis if the wires in the grid are too far apart or too small in diameter to provide adequate support for the feet.

Specialized types of caging that serve specific functions are available for rodents, including caging designed to collect excreta, monitor physiologic characteristics, test behavioral responses, control aspects of the physical environment, and permit enhanced microbiologic control of the environment. Such caging can pose special cleaning and sanitation problems.

Various racking systems, both fixed and mobile, are available to hold either solid-bottom or wire-bottom cages. Racks should be constructed of durable, smooth-surfaced, nonporous materials that can be easily sanitized. Mobile racks are most commonly used because they allow greater flexibility of room arrangement and are easier to clean than fixed racks. If fixed racks are used, adequate steps should be taken to protect floors or walls from damage caused by the weight of the racks and to provide for cleaning under and between the racks. Some racks incorporate devices that automatically supply water directly to the cages they hold.

Housing Systems

Many types of housing systems with specialized caging and ventilation equipment are available for rodents. Generally, the purpose of these housing systems is to minimize the spread of airborne microorganisms between cages; but they often do not prevent transmission of nonairborne fomites. The most frequently used of these systems is the filter-top cage, which has a spun-bound or woven synthetic filter that covers the wire-mesh top of a solid-bottom cage, thereby preventing the entry or escape of airborne particles that can act as fomites for unwanted microorganisms. The use of filter tops restricts ventilation and can alter the microenvironment of the rodents housed in the cages; therefore, to maintain a healthful environment, it might be necessary to change the bedding and clean the cages more often (Keller et al., 1989).

A cubicle (also called an Illinois cubicle or a cubical containment system) is an enclosed area of a room capable of housing one or more racks of cages. It is separated from the rest of the room by a door that usually opens and closes vertically. The cubicle is supplied by air that moves under the door from the room and is exhausted through the ceiling, or a separate air supply is provided to the cubicle through an opening in a wall, the base, or the ceiling. Cubicles have been used to reduce airborne cross contamination between groups of animals housed in conventional plastic or wire-bottom cages (White et al., 1983). They provide better ventilation than many housing methods, but they do not protect against fomite transmission of microorganisms. Strict adherence to sanitation and other husbandry procedures is required if cubicles are to be used effectively.

In some housing systems, cages are individually ventilated with highly filtered air. In some, exhaust air is also filtered or controlled in a way that greatly minimizes the risk of contaminating animals in other cages and personnel in the animal rooms. Such systems can overcome the disadvantages of using nonventilated filter-topped cages while minimizing airborne cross-contamination.

A housing system that is particularly useful for maintaining the microbiologic status of rodents has isolators made of rigid or flexible-film plastic that are designed to enclose a group of rodent cages. Built-in gloves allow the manipulation of animals and materials in the isolators. Isolators are supplied with filtered air and have a filtered exhaust; at least one transfer device is provided for moving sterilized or disinfected materials into the isolator. To maintain the microbiologic status of an isolated group of animals, it is necessary to sterilize or otherwise disinfect all the interior surfaces of the isolators, and all materials introduced into the isolators should be first sterilized or otherwise disinfected.

Space Recommendations

It is generally assumed that there are critical measures of cage floor area and cage height below which the physiology and behavior of laboratory rodents will be adversely affected, thereby affecting the well-being of the animals and potentially influencing research outcomes. However, there are very few objective data for determining what those critical measures are or even whether such interactions exist. A number of studies designed to evaluate the effects of space on population dynamics have been conducted on wild and laboratory rodents housed in a laboratory environment (e.g., see Barnett, 1955; Christian and LeMunyan, 1958), but some of them used caging systems different from those generally used in laboratory animal facilities (e.g., see Davis, 1958; Joasoo and McKenzie, 1976; Thiessen, 1964). Changes in behavior, reproductive performance, adrenal weights, glucocorticoid and catecholamine concentrations, immunologic function, numbers of some kinds of white blood cells (usually lymphocytes), and cage-use patterns have been assessed in those studies and suggested as indicators of stress and compromised well-being (e.g., see Barrett and Stockham, 1963; Bell et al., 1971; Christian, 1960; Poole and Morgan, 1976; White et al., 1989). However, there has never been general agreement as to which physiologic and behavioral characteristics are indicative of well-being in rodents or what magnitude of change in them would be necessary to compromise the well-being of the animals.

With few objective data available, cage space recommendations have been based on the results of surveys of existing conditions and professional judgment and consensus. The *Guide* (NRC, 1996 et seq.) provides space recommendations for rodents. Space recommendations have also been developed in other countries (CCAC, 1980; Council of Europe, 1990), but they are not totally compatible with those in the *Guide*. It is important to remember that space recommendations in the *Guide* serve only as a starting point for determining space required by rodents and might need adjustment to fit the needs of the animals and the purposes for which they are housed.

Although comprehensive studies involving all the characteristics associated with housing rodents are not available, sufficient information does exist to suggest that individually housed rodents and group-housed rodents have different space requirements. For the most part, laboratory rodents are social animals and probably benefit from living in compatible groups (Brain and Bention, 1979; NRC, 1978; White, 1990). Although more study is needed, rodents maintained for long periods, as in lifetime studies, appear to survive longer when housed in large, compatible social groups than when housed in small groups or individually (Hughes and Nowak, 1973; Rao, 1990). Individual housing is sometimes necessitated by the nature of the experimental protocol; in such instances, adequate space should be allotted to allow the animals to make normal postural adjustments, which will depend on the body size attained by the animals during the course of the experiment. Under those circumstances, current space guidelines might not be sufficient, especially if an animal's size exceeds the scope of the recommendations.

Conversely, group-housed rodents would be expected to need less space per animal than individually housed rodents because each animal can also use the space of the other rodents with which it is housed. Studies have found that compatible social groups of rodents do not use all the available space recommended in current guidelines and probably do not require it for well-being (White, 1990; White et al., 1989). Rodents housed in compatible groups share cage space by huddling together along walls and under overhanging portions of the cage, such as feeders, as well as piling up on top of each other during long rest periods. The center of the cage is used infrequently.

Even if individually housed, rodents appear to prefer sheltered areas of the cage, especially if those areas have decreased light and height. Providing such a confined space within a cage might be one way to enrich the environment of rodents.

Sexually mature male rodents often fight when housed in groups for breeding or other purposes, but this behavior has never been shown to be a function of the amount of available floor space in the cage. Rather, the incidence of fighting appears to be related more to combining males into groups when they are sexually mature (especially if females are housed in the same room) or have participated in mating programs. Increasing the cage space is not effective in preventing the development of such behavior or in eliminating it once it has occurred. Only separation of the animals into individual cages or into smaller, compatible groups is effective in eliminating fighting.

In determining adequate cage space, it is important to consider the conditions of the experimental procedure and how long the animals will be housed. Animals that become debilitated during the course of an experimental procedure might require increased cage space or an alteration in caging to accommodate limitations in motion, recumbent positions, and the need for alternative food and water sources. Older animals are less active than younger animals and use less of the cage space or available activity devices.

The *Guide* (NRC, 1996 et seq.) and other guidelines also recommend cage heights. The recommendations do not appear to be related to the body size of rodents nor to their normal locomotion patterns. Laboratory rodents exhibit some vertical exploratory behavior when put into a new cage, and it has been suggested that relatively high cages be provided to accommodate this occasional behavior (Lawlor, 1990; Scharmann, 1991). However, there is no good evidence to suggest that rodents require tall enclosures. On the contrary, as described previously, they tend to seek shelter under objects lower than recommended in existing guidelines. Depending on the caging type, existing height guidelines can be useful for ensuring that there is adequate space for side-wall or cage-top feeders and adequate clearance for sipper tubes or other watering devices.

In summary, the space required to maintain rodents, either individually or in groups, depends on a number of factors, including age, weight, body size, sexual maturity, experimental intervention, behavioral characteristics, the duration of housing, group size, breeding activities, and availability of enrichment devices or sheltering areas within the cage. The relationships among those factors are complex, and there is not necessarily a direct correlation between body weight or surface area of the animals and the absolute floor area of the cage required or used by them. Guidelines should be used with professional judgment based on assessment of the animals' well-being. However, alterations that bring floor area or height of cages below recommended levels should be adequately justified and approved by the IACUC.

ENVIRONMENT

Microenvironment

The microenvironment of a rodent is the physical environment that immediately surrounds it and is usually considered to be bounded by the primary enclosure or cage in which it resides. In contrast, the physical conditions in the secondary enclosure or animal room make up the macroenvironment. The components of the macroenvironment are often easier to measure and characterize than those of the microenvironment. The two environments are linked or coupled, but the character of each is often quite different and depends on a variety of factors, such as the numbers and species of rodents housed in the microenvironment, the design and construction of the cages, and the types of bedding materials used (Besch, 1975; Woods, 1975; Woods et al., 1975).

The measurement of constituents of the microenvironment of rodents is often difficult because of the relatively small volume of the primary enclosure. Available data show that temperature, humidity, and concentrations of gases and particulate matter—such as carbon dioxide, ammonia, methane, sulfur dioxide, respirable particles, and bacteria—are often higher in the microenvironment than in the macroenvironment (Besch, 1980; Clough, 1976; Flynn, 1968; Gamble and Clough, 1976; Murakami, 1971; Serrano, 1971). Although there is little information on the relation between the magnitude of exposure to some of those components and alterations in disease susceptibility or changes in metabolic or physiologic processes, the available data clearly suggest that the characteristics of the microenvironment can have a substantial impact on research results (Broderson et al., 1976; Vessell et al., 1973, 1976).

Temperature

Temperature and relative humidity are important components of the environment of all animals because they directly affect an animal's ability to regulate internal heat. They act synergistically to affect heat loss in rodents, which lose heat by insensible means, rather than by perspiring. Studies in the older literature, which were conducted without the benefit of modern systems for controlling conditions precisely or modern instrumentation, have established that extremes in temperature can cause harmful effects (Lee, 1942; Mills, 1945; Mills and Schmidt, 1942; Ogle, 1934; Sunstroem, 1927). However, those studies were done on only a few laboratory species.

Studies in the past generally focused on prolonged exposure of laboratory animals to temperatures above 85°F (29.4°C) or below 40°F (4.4°C), which are required to achieve clinical effects (Baetjer, 1968; Mills, 1945; Weihe, 1965). When exposed to those extreme temperatures, rodents use behavioral means (e.g., nest-building, curling up, huddling with others in the cage, and adjusting activity level) to attempt to adapt. If the temperature change is brief or small, behavioral adaptation is sufficient; profound or prolonged temperature changes generally require physiologic or structural adaptation as well. Physiologic adaptation includes alterations in metabolic rate, growth rate, and food or water consumption; hibernation or estivation; and the initiation of nonshivering thermogenesis. Structural adaptation includes alterations in fat stores, density of the hair coat, and structure or perfusion of heat-radiating tissues and organs (e.g., tail, ears, scrotum, and soles of the feet). Initiation of such changes usually requires exposure to an extreme temperature for at least 14 days.

For routine housing of laboratory rodents, a consistent temperature range should be provided. However, there is little scientific evidence from which optimal temperature ranges for laboratory rodents can be determined. For each species, there is a narrow range of environmental temperatures at which oxygen consumption is minimal and virtually independent of change in ambient temperature. The range in which little energy is expended to maintain body temperature is called the thermal neutral zone, and some have suggested that it is a range of comfortable temperatures for rodents (Besch, 1985; Weihe, 1965, 1976a). However, other evidence suggests that animals held within this temperature range do not necessarily achieve optimal growth and reproductive performance, and the optimal temperature range might be age-dependent (Blackmore, 1970; Weihe, 1965). Moreover, measurements of thermal neutral zones are generally made on resting animals and do not take into account periods of increased activity or altered metabolic states, such as pregnancy. Thermal neutrality does not necessarily equate with comfort. In the absence of well-controlled studies that used objective measures for determining optimal ranges, recommended temperature ranges for laboratory rodents have been independently developed by several groups on the basis of professional judgment and experience (e.g., CCAC, 1980; Council of Europe, 1990; NRC, 1996 et seq.).

Humidity

Relative humidity varies considerably with husbandry and caging practices. In addition, there is usually a difference between the relative humidity in the room and that in the animal cages. Many factors—including cage material and construction, use of filter tops, number of animals per cage, frequency of bedding changes, and bedding type—can affect the relative humidity in the rodents' immediate environment.

Variations in relative humidity appear to be tolerated much better at some temperatures than at others. Studies in humans and limited in vitro work on survival of microorganisms have established a loose association between humidity and susceptibility to disease (Baetjer, 1968; Dunklin and Puck, 1948; Green, 1974; Webb et al., 1963), but there is no good evidence to establish this link in animals. Low relative humidity has been reported to be associated with the development of "ring tail" in rodents (Flynn, 1959; Njaa et al., 1957; Stuhlman and Wagner, 1971); however, this condition has not been adequately studied and does not appear to be reproducible by lowering relative humidity in controlled laboratory experiments.

Guidelines have been established for relative-humidity ranges based on experience and professional judgment (CCAC, 1980; Council of Europe, 1990; NRC, 1996 et seq.). There is no evidence to support limiting the variation of relative humidity within these ranges; however, the combination of high relative humidity and high environmental temperature can affect the ability of rodents to dissipate heat by insensible means and should be avoided.

Ventilation

Ventilation Rate

Ventilation refers to the process of using conditioned air to affect temperature, humidity, and concentrations of gaseous and particulate contaminants in the environment. Ventilation is often characterized at the animal-room level as air exchanges per hour. However, as for other environmental conditions, there are no definitive data showing that the air-exchange range in existing guidelines (i.e., 10-15 air changes/hour) provides optimal ventilation for laboratory rodents.

Existing guidelines have been criticized as being based mainly on keeping odors below objectionable limits for humans (Besch, 1980; Runkle, 1964) and, in recent years, as being energy-intensive. An often-quoted study by Munkelt (1938) appears to support the first contention: his measure of adequate ventilation was the ability to smell ammonia in the environment. Besch (1980) suggested that ventilation should be based on air-exchange rate per animal or animal cage because room air-exchange rates do not consider such factors as population density, room configuration, and cage placement within a room. Ultimately, however, the ventilation rate in animal rooms is governed by the heat loads produced in the rooms, which include not only heat produced by animals but also that produced by other heat-radiating devices, such as lighting (Curd, 1976).

Available evidence suggests that little additional control of the concentrations of gaseous and particulate contaminants is gained by using air-exchange rates higher than those recommended in current guidelines (Barkley, 1978; Besch, 1980). The recommendation of providing a room air-exchange rate of 10-15 changes/hour is still useful; however, this ventilation range might not be appropriate in some circumstances, especially if the diffusion of air within the room is inappropriate for the type and placement of cages. Other methods of providing equal or more effective ventilation, including the use of individually ventilated cages or enclosures and the adjustment of ventilation rate to accommodate unusual population densities, are also acceptable.

Calculation of the amount of cooling required to control expected sensible and latent heat loads generated by the species to be housed and the largest expected population (ASHRAE, 1993) can be used to determine minimal ventilation requirements. However, that calculation does not take into account the generation of odors, particles, and gases, which might necessitate greater ventilation.

Air Quality

The quality of air used to ventilate animal rooms is another important consideration. Ventilation systems for rodent rooms incorporate various types of filtration of incoming air. Coarse filtration of the air supply is a minimal requirement for proper operation of ventilating equipment. Most facilities maintaining rodents of defined microbiologic status also use high-efficiency particulate air (commonly called HEPA) filters to decrease the risk of introducing rodent pathogens into the animal room through the fresh-air supply (Dyment, 1976; Harstad et al., 1967). The required filter efficiency is a matter of professional judgment, and selection should be based on the perceived likelihood of introducing contaminated air into the room. Filtration of exhaust air from rodent rooms when air is not recycled is usually deemed unnecessary unless the exhaust air is likely to contain hazardous or infectious materials. Filters designed to remove chemicals from air are sometimes incorporated into exhaust systems to remove animal odors. Activated-chemical filters (e.g., those with activated charcoal) are often used for this purpose; however, their efficiency in removing odoriferous compounds, including ammonia, varies, and they require substantial maintenance to remain effective.

The use of recycled air to ventilate animal rooms can save considerable amounts of energy. However, many animal pathogens can be airborne or travel on fomites, such as dust, so recycling of exhaust air into heating, ventilating, and air-conditioning systems that serve multiple rooms presents a risk of cross contamination. Exhaust air that is to be recycled should be HEPA-filtered to remove particles. HEPA filters are available in various efficiencies; the extent and efficiency of filtration should be proportional to the risk. Toxic or odor-causing gases produced by decomposition of animal wastes can be removed by the ventilating system with chemical absorption or scrubbing, but those methods might not be completely effective. Frequent bedding changes and cage-cleaning, a reduction in number of animals housed in a room, and a decrease in environmental temperature and humidity—within limits recommended in the *Guide* (NRC, 1996 et seq.)—can also assist in reducing the concentration of toxic or odor-causing gases. Treatment of recycled air to remove either particulate or gaseous contaminants is expensive and can be ineffective if filtration systems are improperly or insufficiently maintained. Therefore, recycling systems require regular monitoring for effective use.

An energy-recovery system that reclaims heat and thereby makes it energy-efficient to exhaust animal-room air totally to the outside is also acceptable, but these systems often require much maintenance to be effective. The recycling of air from nonanimal areas can be considered as an alternative to the recycling of animal-room air, but this air might require filtering and treatment to remove odors, toxic chemicals, and particles (White, 1982).

Relative Air Pressures

To minimize the potential for airborne cross-contamination between adjacent rodent rooms or between rodent rooms and other areas where contaminants might be generated, it is important to consider controlling relative air pressures. By adjusting the rates of air flow to and from individual areas, one can produce a negative or positive pressure relative to adjoining areas. When the intent is to contain contaminants (e.g., in areas used to quarantine newly arrived animals, isolate animals infected or suspected of being infected with rodent pathogens, house animals or materials inoculated with biohazardous substances, or keep soiled equipment), air pressure in the containment area should be lower than that in surrounding areas. When the intent is to prevent the entry of contaminants, as in areas used to house specific-pathogen-free rodents or keep clean equipment, air pressure in the controlled area should be greater than that in surrounding areas. It is important to remember, however, that many factors influence disease transmission between adjacent rooms; simply controlling air pressure is not sufficient to prevent transmission.

Cage Ventilation

Ventilation can easily be measured in rodent-holding rooms; however, conditions monitored in a room do not necessarily reflect conditions in the cages in the room. The large sample volumes required by the commonly used instruments that measure ventilation, as well as the size of the intruments themselves, preclude accurate measurement in cages (Johnstone and Scholes, 1976). The degree to which cages are ventilated by the room air supply is affected by cage design; room air-diffuser type and location; number, size, and type of animals in the cages; presence of filter tops; and location of the cages. For example, cages without filter tops provide better air and heat exchange than those with filter tops, in which ventilation is substantially decreased (Keller et al., 1989). Rigidly maintaining room air quality and ventilation will not necessarily provide the same environment for similar groups of animals or even for similar cages in the same room. Individually ventilated cages provide better ventilation for the animals and, possibly, a more consistent environment (Lipman et al., 1992), but those systems are generally expensive.

It has not been established whether rodents are uncomfortable when exposed to air movements (drafts) or that exposure to drafts has biologic consequences. However, movement of air in a room influences the ventilation of an animal's primary enclosure and so is an important determinant of microenvironment.

Illumination

Animal-room lighting can affect the eyes of laboratory rodents, especially albino rodents. In examining the effects, there is a tendency to think only in terms of light intensity. However, it is the interaction of the three characteristics of light (spectral distribution, photoperiod, and intensity) that produces the effects (Brainard, 1988; Wurtman et al., 1985. Also contributing to the effects of light exposure is the amount of time that rodents have their eyes open during the hours when the room is lit. Those factors should be kept in mind in reading the following discussion.

Spectral Distribution

Artificial lighting with white incandescent or fluorescent fixtures is preferred for rodent housing facilities because it provides consistent illumination. The two types of lighting have similar spectra, although incandescent lighting generally has more energy in the red wavelengths and less energy in the blue and ultraviolet (UV) wavelengths than white fluorescent lighting. Although some fluorescent lighting more closely simulates the wavelength distribution of sunlight than incandescent lighting, no artificial lighting truly duplicates sunlight, and there is little reason to believe that the spectral distribution of one type of artificial lighting is superior to that of any other for rodent rooms. There is some evidence that UV light can increase the incidence of cataract formation in humans (Zigman et al., 1979) and in rodents exposed to very high levels (Zigman and Vaughan, 1974; Zigman et al., 1973). However, there is no evidence that UV-associated cataracts develop in rodents maintained under levels of illumination normally found in animal rooms. UV radiation from fluorescent lights is eliminated when the lights are covered by plastic diffusing screens (Kaufman, 1987; Thorington, 1985).

Photoperiod

Photoperiod (cycles of light and dark during the course of a single day) affects various physiologic and metabolic characteristics, including reproductive cycles, behavioral activity, and the release of hormones into the blood (Brainard, 1989; Reiter, 1991). The rods and cones in the eye are influenced by photoperiod, requiring an interval of darkness for regeneration (LaVail, 1976; Williams, 1989; Williams and Baker, 1989). There is evidence that exposure to even low-intensity light—64.6-193.7 lx (6-18 ft-candles)—continuously for 4 days can cause degenerative retinal changes (Anderson et al., 1972; O'Steen, 1970; Williams, 1989).

Photoperiods in rodent rooms are usually controlled by automatic timers. The cycles usually recommended are either 12 hours of light and 12 hours of dark or 14 hours of light and 10 hours of dark. For some mammals (e.g., hamsters), a longer period of light is important for normal reproduction (Alleva et al., 1968). In general, lighting in laboratory animal facilities does not reproduce that in nature, in that most light-timing devices do not provide any interval of reduced lighting intensity (simulating dawn and dusk). Changes or interruptions in light-dark cycles should be avoided because of the importance of photoperiod in normal rodent reproduction and other light-affected physiologic processes (Weihe, 1976b). Similarly, light from exterior windows and uncontrolled hallway lighting are usually undesirable.

Light-timing devices in rodent facilities should be checked regularly for correct operation. Any system that can be overridden manually should be equipped with an indicator, such as a light, to remind personnel to turn off the override device or with a timer to turn it off automatically. Photoperiod can also be checked by photosensors linked to a computer-based monitor.

Intensity

The intensity of illumination is inversely proportional to the square of the distance from the source. Therefore, statements about intensity should indicate where it was measured. In animal facilities, such statements generally specify distance above the floor; that implies that the illumination is uniformly diffused throughout the room. The actual intensity experienced by a rodent in an animal room is influenced not only by the relative locations of its cage and the room lights, but also by cage material and design.

The optimal light intensity required to maintain normal physiology and good health of laboratory rodents is not known. In the past, illumination of 807-1076 lx (75-100 ft-candles) or higher has been recommended to allow adequate observation of the animals and safe husbandry practices (NRC, 1978). The point of measurement for that recommendation was never clearly stated, but it has been generally assumed that the recommendation referred to the illumination at maximal cage height in the center of the room. The recommended intensities, however, have been shown to cause retinal damage in albino mice (Greenman et al., 1982) and rats (Lai et al., 1978; Stotzer et al., 1970; Williams and Baker, 1980).

More recently, a light intensity of 323 lx (30 ft-candles) measured about 1.0 m (3.3 ft) above the floor has been recommended as adequate for routine animal care (Bellhorn, 1980; NRC, 1996 et seq.). That intensity has been calculated to provide 32-40 lx (3.0-3.7 ft-candles) to rodents in the front of a cage that is in the upper portion of a cage rack. Exposure for up to 90 days to an intensity of around 300 lx during the light cycle has been reported not to cause retinal lesions in rats (Stotzer et al., 1970); however, it is still questionable whether exposure to light of even this intensity can cause retinal lesions in albino animals if they are exposed for longer periods (Weisse et al., 1974).

Alternatives to providing a single light intensity in a room are to use variable-intensity controls and to divide rooms into zones, each lighted by a separate switching mechanism. Those alternatives conserve energy and provide sufficient illumination for personnel to perform their tasks adequately and safely. However, caution is necessary when instituting those alternatives. Boosting daytime room illumination for maintenance purposes has been shown to change photoreceptor physiology and can alter circadian regulation (Remé et al., 1991; Society for Research on Biological Rhythms, 1993; Terman et al., 1991).

Noise

Many sounds of varied frequencies and intensities are generated in animal facilities during normal operation. Rodents emit ultrasonic vocalizations that are an important part of their social and sexual behavior. Rats can hear frequencies as high as about 60-80 kHz but are relatively insensitive to frequencies less than 500 Hz (Kelly and Masterton, 1977; Peterson, 1980). Sounds are also produced by mechanical equipment (less than 500 Hz): by dog, cat, nonhuman primate, and pig vocalizations (up to 120 dB at 500 Hz); and by exterior conditions (e.g., highway noise).

If acoustic energy is high enough (80-100 dB), both auditory and nonauditory changes can be detected in laboratory animals (Algers et al., 1978; Moller, 1978). The type of change produced depends on the pattern of sound presentation. Sound of uniform frequency and unchanging intensity can cause hearing loss in some rodents (Bock and Saunders, 1977; Burdick et al., 1978; Kelly and Masterton, 1977; Kraak and Hofmann, 1977). Hamsters, guinea pigs, rats, and mice pass through developmental stages during which they are very susceptible to injury from sound of this type (Kelly and Masterton, 1977). Sound of irregular frequency and rapidly changing intensity that is presented to animals in an unpredictable fashion can cause stress-induced mechanical and metabolic changes (Anthony and Harclerode, 1959; Geber, 1973; Guha et al., 1976; Kimmel et al., 1976; Peterson et al., 1981). Continuous exposure to acoustic energy above 85 dB can cause eosinopenia (Geber et al., 1966; Nayfield and Besch, 1981), increased adrenal weights (Geber et al., 1966; Nayfield and Besch, 1981), and reduced fertility (Zondek and Tamari, 1964).

Few studies are available on the long-term effects on rodents of sound comparable with that normally encountered in rodent rooms, and there are hardly any data on the sensitivity of rodents to intensity as a function of frequency (Peterson, 1980). In addition, no comparative damage-risk criteria have been established for rodents; therefore, recommendations for acceptable noise in animal facilities are often based on extrapolations from humans (Peterson, 1980). As a general guideline, an effort should be made to separate rodent-housing areas from human-use areas, especially human-use areas where mechanical equipment is used or where noisy operations are conducted. Common soundproofing materials are not compatible with some of the construction requirements for animal facilities designed to house rodents, but attention can be given to separating rooms housing rodents from those housing noisy species, such as nonhuman primates, dogs, cats, and swine. The location of loud, unpredictable sources of noise—such as intercoms, paging systems, telephones, radios, and alarms—should be carefully considered because the noise from such sources can be stressful to some rodents. Extra care should be taken to control noise in rooms that house rodents that are subject to audiogenic seizures. Every reasonable effort should be made to house rodents in areas away from environmental sources of noise.

FOOD

Nutrition has a major influence on the growth, reproduction, health, and longevity of laboratory rodents, including their ability to resist pathogens and other environmental stresses and their susceptibility to neoplastic and nonneoplastic lesions. Providing nutritionally adequate diets is important not only for the rodents' welfare, but also to ensure that experimental results are not biased by unintended or unknown nutritional factors. Providing nutritionally adequate diets for laboratory rodents involves establishing requirements for about 50 essential dietary nutrients, formulating and manufacturing diets with the required nutrient concentrations, and managing numerous factors related to diet quality. Factors that potentially affect diet quality include bioavailability of nutrients, palatability or acceptance by the animals, preparation and storage procedures, and concentrations of chemical contaminants. The estimated nutrient requirements of laboratory animal species are periodically reviewed and updated by a committee of the National Research Council (NRC, 1995), and information about the formulation, manufacture, and management of laboratory animal diets is available elsewhere (Coates, 1987; Knapka, 1983, 1985; McEllhiney, 1985; Navia, 1977; Rao and Knapka, 1987).

Types of Diets

Adequate nutrition can be provided for laboratory rodents in different types of diets that are classified by the degree of refinement of the ingredients used in their formulation (NRC, 1995).

Natural-ingredient diets are formulated with agricultural products and byproducts, such as whole grains (e.g., ground corn and ground wheat), mill byproducts (e.g., wheat bran, wheat middlings, and corn gluten meal), high-protein meals (e.g., soybean meal and fish meal), processed mineral sources (e.g., bone meal), and other livestock feed ingredients (e.g., dried molasses and alfalfa meal). Commercial diets are the most commonly used natural-ingredient diets, but special diets for specific research programs can also be of this type if appropriate attention is given to ingredient selection and diet formulation. Natural-ingredient diets are relatively inexpensive to manufacture and are readily consumed by laboratory rodents.

A natural-ingredient diet can be either an open-formula diet (information on the amount of each ingredient in the diet is readily available) or a closed-formula diet (information on the amount of each ingredient is privileged). The advantages of using natural-ingredient, open-formula diets in biomedical research have been discussed (Knapka et al., 1974).

There are two concerns about the use of natural-ingredient diets in biomedical research. First, such factors as varieties of plants, soil compositions, weather conditions, harvesting and storage procedures, and manufacturing and milling methods influence the nutrient composition of ingredients used in this type of diet to the extent that no two production batches of the same diet are identical (Knapka, 1983). This variation in dietary-nutrient concentrations introduces an uncontrolled variable that could affect experimental results. Second, natural ingredients can be exposed to various naturally occurring or human-made contaminants, such as pesticide residues, heavy metals, and estrogen. Diets manufactured from natural ingredients can contain low concentrations of contaminants that might have no influence on animal health but could affect experimental results. For example, a lead concentration of 0.5-1 part per million is inherent in natural-ingredient rodent diets and is not generally detrimental to animal health; but it could substantially influence the results of toxicology studies designed to evaluate lead-containing test compounds.

Purified diets are formulated with ingredients that have been refined so that in effect each ingredient contains a single nutrient or nutrient class. Examples of the ingredients are casein or soy protein isolate, which provide protein and amino acids; sugar and starch, which provide carbohydrates; vegetable oil and lard, which provide essential fatty acids and fat; a chemically extracted form of cellulose, which provides fiber; and chemically pure inorganic salts and vitamins. The nutrient concentrations in this type of diet are less variable and more readily controlled than those in natural-ingredient diets. However, purified ingredients can contain low and variable concentrations of trace minerals, and batch-to-batch variation in their concentrations is inherent in the manufacture of purified diets. The potential for chemical contamination of purified diets is low; however, they are not always readily consumed by laboratory rodents, and they are more expensive to produce than natural-ingredient diets.

Chemically defined diets are formulated with the most elemental ingredients available, such as individual amino acids, specific sugars, chemically defined triglycerides, essential fatty acids, inorganic salts, and pure vitamins. Use of this type of diet provides the highest degree of control over dietary nutrient concentrations. However, chemically defined diets are not readily consumed by laboratory rodents, and they are usually too expensive for general use.

The dietary nutrient concentrations in chemically defined diets are theoretically fixed at the time of manufacture; however, the bioavailability of nutrients can be altered by oxidation or nutrient interactions during diet storage. Liquid chemically defined diets that can be sterilized by filtration have been developed (Pleasants, 1984; Pleasants et al., 1986).

Criteria for Selecting Optimal Rations

Selection of the most appropriate type of diet for a particular animal colony depends on the reproductive or experimental objectives. One of the most important considerations is the amount of control over dietary-nutrient composition that is necessary to attain the objectives. For example, the use of a purified diet is essential for studies designed to establish quantitative requirements for micronutrients because the batch-to-batch variation in nutrient concentrations inherent in natural-ingredient diets would compromise experimental results. Conversely, the variation in nutrient concentrations in natural-ingredient diets would have no detectable influence on rodent production colonies because the nutrient concentrations are generally greater than those required in a nutritionally adequate diet. The use of chemically defined diets might be required for studies whose objectives involve dietary concentrations of single amino or fatty acids.

The potential for chemical contamination is an important consideration in selecting a diet for rodents that will be used in toxicology studies. Even though the concentrations of chemical contaminants in natural-ingredient diets are so low that they generally do not jeopardize animal health, they might be high enough to compromise results of toxicology studies. The results of some immunology studies might also be influenced by the use of natural-ingredient diets because some ingredients, particularly those of animal origin, contain antigens. Purified diets should be considered for animals used in both kinds of studies, although their cost can increase the cost of conducting the research, especially in life-span studies that use large numbers of rodents.

Any diet selected should be accepted by the animals, otherwise considerable amounts will be wasted. This is expensive and constitutes a major disadvantage in studies that require quantification of dietary consumption. Diets should be nutritionally balanced and free of toxic or infectious agents. If diet is a factor in a study, the diet selected should be readily reproducible to ensure that research results can be verified by replication.

Quality Assurance

Although reputable laboratory animal feed manufacturers develop elaborate programs to ensure the production of high-quality products, additional procedures are often required to ensure that the diets are nutritionally adequate. The shelf life of any particular feed lot depends on the environmental conditions during storage. Nutrient stability of animal feeds generally increases as temperature and humidity in the storage environment decrease. Natural-ingredient rodent diets stored in air-conditioned areas in which the temperature is maintained below 21°C (70°F) and the humidity below 60 percent should be used within 180 days of manufacture. Vitamin C in diets stored under these conditions has a shelflife of only 90 days. If a vitamin C-containing diet stored for more than 90 days is to be fed to guinea pigs, an appropriate vitamin supplement should be added. To monitor compliance with these guidelines, storage containers should be marked with the date of manufacture of the food stored therein.

Diets stored for longer periods or under conditions other than those recommended above should be assayed for the most labile nutrients (i.e., vitamin A, thiamine, and vitamin C) before use. Diets formulated without antioxidants or with large amounts of highly perishable ingredients, such as fat, might require special handling or storage procedures.

Given the potential importance of diet quality and consistency to experimental results, a routine program of nutrient testing should be implemented to verify the composition of diets fed to research animals. Accidental omission or inclusion of ingredients in the manufacturing process, although uncommon, can have disastrous consequences on research projects. Discrepancies between expected and actual nutrient concentrations in laboratory animal diets can arise from errors in formulation, which can result in hazardous concentrations of nutrients that are toxic when present in excess of requirements (e.g., vitamins A and D, copper, and selenium); losses of labile nutrients during manufacture or storage; variation in nutrient content of ingredients used in diet formulation; and errors associated with diet sampling or analysis. Although most laboratory animal feed manufacturers will provide data on the complete nutrient composition of rodent diets, it is often difficult to ascertain the source of these data (i.e., whether they are calculated, representative of several diet production batches, or representative of a single production batch). Therefore, it is suggested that feed manufacturers routinely be asked to provide the results of nutrient assays of representative samples of their diets.

Testing samples of natural-ingredient diets used in research colonies is particularly important because the nutrient concentrations measured by analysis can differ from the expected concentrations. Samples for assay should be collected from multiple bags or containers within a single production batch of feed (i.e., in which all containers bear the same manufacture date). The containers sampled should be selected at random; traditionally, the number sampled equals the square root of the total number of containers in a single shipment or production batch. The objective is to obtain a sample of diet that is representative of the entire lot being assayed. Nutrient analyses should be conducted by a laboratory with an established reputation in assaying feed samples, and all assays should be conducted in accordance with the most recent methods published by the Association of Official Analytical Chemists (Helrich, 1990). Analyses should include at least proximate constituents (i.e., moisture, crude protein, ether extract, ash, and crude fiber) and any nutrients that are under study or that could influence the study. Some vitamins and other nutrients required at trace concentrations might be difficult to assay because of low concentrations, interfering compounds, or both.

The presence of biologic contaminants in diets is a cause for concern in most research and production rodent colonies. Unwanted agents in the diet include pathogenic bacteria and viruses, insects, and mites. Diets for axenic and microbiologically associated rodents should be sterilized before use, as should those for severely immunodeficient rodents (i.e., athymic rodents and mice homozygous for the mutation *scid*) (NRC, 1989). Diets for specific-pathogen-free (SPF) rodents should be subjected to some degree of decontamination, such as pasteurization. It is also prudent to decontaminate diets, at least partially, for conventionally maintained rodents, particularly when they are used in long-term studies. Steam autoclaving is the most widely used method for eliminating biologic contaminants from diets (Coates, 1987; Foster et al., 1964; Williams et al., 1968). However, this process can decrease the concentrations of heat-labile nutrients (Zimmerman and Wostmann, 1963). To ensure that adequate amounts of the most heat-labile vitamins (e.g., vitamins A and C and some of the B complex) will remain after autoclaving, consideration should be given to purchasing autoclavable diets that have been fortified with those vitamins. The magnitude of fortification in autoclavable diets is not generally high enough to be toxic to rodents; however, the routine use of autoclavable diets without autoclaving is not recommended, because the increased vitamin concentrations could influence experimental results.

The level of sterility required for axenic or microbiologically associated rodents requires that the temperature of the diet be raised above 100°C (212°F). To ensure that all the diet in the autoclave attains this temperature, it is recommended that the diet be exposed to a temperature of 121°C (250°F) for 15-20 minutes. Diets should not be subjected to the maximal autoclaving temperature longer than necessary to achieve sterilization (Coates, 1987).

To ensure proper operation of the autoclave, sterility of the diet, and adequate concentrations of labile nutrients, validation procedures are required, including periodic evaluation of autoclave operation by qualified personnel, use of commercially available heat indicators, culture of autoclaved feed samples for biologic contaminants, and assay of autoclaved feed samples to verify nutritional adequacy. Clarke et al. (1977) have described procedures for sampling and assaying feeds for various pathogenic organisms and provided standards for the number and kinds of organisms that are acceptable in diets.

Autoclaving at 80°C (176°F) for 5-10 min is required for pasteurization of diets. At that temperature, vegetative forms, but not spores, of microorganisms are destroyed (Coates, 1987). Pasteurized diets are generally acceptable for use in both specific-pathogen-free and conventional rodent colonies. Pasteurization, rather than sterilization, is used because there is less nutrient loss, and the diets are more readily consumed than are sterilized diets.

Laboratory rodent diets also can be decontaminated by ionizing radiation (Coates, 1987; Coates et al., 1969; Ley et al., 1969), and diets sterilized in this way are now commercially available. Ethylene oxide fumigation has also been used to decontaminate diets (Meier and Hoag, 1966).

All animal diets, particularly those produced from natural ingredients, can contain or become contaminated with various manufactured or naturally occurring chemicals, including pesticide residues, bacterial or plant toxins, mycotoxins, nitrates, nitrites, nitrosamines, and heavy metals (Fox et al., 1976; Newberne, 1975; Yang et al., 1976). Procedures, if any, for detecting these chemicals are often difficult and expensive. Testing for contaminant concentrations in natural-ingredient diets should be routine in toxicologic research and might be valuable in some other studies.

On the basis of observed contaminant concentrations and potential toxic effects, Rao and Knapka (1987) developed a list of recommended limits for about 40 chemical contaminants. The authors also proposed a scoring system for diets used in chemical toxicology studies that permits separation of tested diets into those acceptable for long-term use, those acceptable only for short-term or transitory use, and those which should be rejected.

Laboratory animal diets designated as "certified" are commercially available. Although the term is subject to different interpretations, in most cases the certification guarantees that the concentration of each contaminant on a specific list will not exceed the indicated maximum. Because the maximal concentrations usually are established by the diet manufacturer, the use of certified diets might not be appropriate for studies in which the acceptable concentrations of contaminants could influence experimental data independently or through an additive effect. In addition, a diet might have contaminants that are not included in the certification but are of concern in specific research projects.

Caloric Restriction

Traditionally, the criterion used to evaluate laboratory rodent diets for nutritional adequacy has been maximal growth or reproduction of the animals consuming the diet. Laboratory rodents generally are given ad libitum access to such diets throughout their lives. However, during the past 60 years, many studies have shown beneficial effects of caloric restriction in various species, including laboratory rodents (Bucci, 1992; Snyder, 1989; Weindruch and Walford, 1988; Yu, 1990). It has been reported that caloric restriction increases life expectancy and life span, decreases the incidence and severity of degenerative diseases, and delays the onset of various neoplasias.

The objective of caloric restriction is to reduce calories without malnourishing the animals. That objective is generally accomplished by supplementing a diet with micronutrients and then limiting dietary consumption to 60-80 percent of the dietary consumption of animals that are fed ad libitum; this procedure results in decreased total caloric consumption. Although studies have been conducted in which the total fat (Iwasaki et al., 1988), protein (Davis et al., 1983; Goodrick, 1978), or carbohydrate (Kubo et al., 1984; Yu et al., 1985) consumption has been limited individually, only reduction in caloric intake results in the full range of dietary-restriction-related beneficial effects. Hypotheses explaining the results of dietary restriction studies have been reviewed and discussed (Keenan et al., 1994).

Numerous questions still need to be addressed to determine by what mechanisms dietary or caloric restriction influences various life processes, and the quantitative nutrient or energy requirements necessary to achieve the effects associated with dietary restriction have not been established. However, the reported data show that ad libitum feeding might not be universally desirable for rodents used in long-term toxicologic or aging studies, and this factor should be a prime consideration when designing such studies.

WATER

Laboratory rodents should have ad libitum access to fresh, potable, uncontaminated drinking water, which can be provided by using water bottles and drinking tubes or an automatic watering system. Occasionally, it is necessary to train animals to use automatic watering devices. If water bottles are used, it is better to replace than to refill them; however, if they are refilled, each bottle should be returned to the cage of origin to minimize potential cross-contamination with microbial agents. If automatic watering devices are used, they should be examined routinely to ensure proper operation. The drinking nozzles on these devices should be sanitized regularly, and the pipe distribution system should be flushed or disinfected routinely.

Water is a potential source of microbial or chemical contaminants. Although a water source might be in compliance with standards that ensure purity of water supplied for human consumption, additional treatment might be required to ensure that water constituents do not compromise animal-colony objectives. Treatments used to limit or eliminate bacteria in water intended for laboratory rodents maintained in axenic or SPF environments include distillation, sterilization by autoclaving, hyperacidification, reverse osmosis, ultraviolet treatment, ultrafiltration, ozonation, halogenation, and irradiation (Bank et al., 1990; Engelbrecht et al., 1980; Fidler, 1977; Green and Stumpf, 1946; Hall et al., 1980; Hann, 1965; Hermann et al., 1982; Kool and Hrubec, 1986; Newell, 1980; Tobin, 1987; Tobin et al., 1981; Wegan, 1982). The advantages, disadvantages, and potential effects of water treatment on an animal's physiologic response to experimental treatments should be evaluated before a method of water decontamination is initiated. In general, any treatment that decreases water consumption is potentially detrimental to the animals' health and welfare.

Drinking water of animals used in toxicology experiments, particularly those of long duration, should be periodically assayed for compounds that might influence experimental results, even when exposures are small. Mineral concentrations in water can have a profound influence on experimental results in studies designed to establish dietary mineral requirements for laboratory rodents. Distilled or deionized drinking water should be provided to rodents used in studies in which the amounts of minerals consumed are critical.

BEDDING

Bedding materials are used to absorb spilled water, minimize urinary and fecal soiling of the animals, and assist in decreasing the generation of odors and gaseous contaminants caused by bacterial decomposition of urine and feces. Bedding material can be used either as contact bedding in solid-bottom cages or as noncontact bedding in waste-collection pans placed beneath wire-bottom cages. Contact bedding provides thermal insulation for the animals and is often used as nesting material in breeding colonies. Abrasive or toxic materials should not be used as contact bedding.

Most products used for bedding in rodent colonies are byproducts of various industries. During the manufacturing process, these byproducts are occasionally subjected to conditions that are conducive to microbial contamination. Many of the commercially available rodent bedding materials are subjected to heat treatment before packaging; however, microbiologic recontamination can occur during shipment from the manufacturing plant to the animal facility. For maximal protection from potential microbiologic contamination, contact and noncontact bedding products should be sterilized before use.

Hardwood and softwood are the most commonly used rodent bedding materials. Wood products should be screened to eliminate splinters or slivers and should be free of foreign materials, such as paint, wood preservatives, chemicals, heavy metals, and pesticides. Some manufacturers will provide an assurance that the bedding is free of specified contaminants. The moisture content of wood products should be high enough to prevent excessive dust but low enough to provide adequate absorbency. Cedarwood products are often mixed with other bedding material to mask animal-room odors; however, their use is not recommended because the aromatic hydrocarbons inherent in these products can alter hepatic microsomal enzyme activity and potentially influence experimental results (Cunliffe-Beamer et al., 1981; Ferguson, 1966; Porter and Lane-Petter, 1965; Vesell, 1967; Vesell et al., 1976). Furthermore, masking animal-room odors with cedar products is not a substitute for good sanitation practices.

Plant byproducts and other cellulose-containing materials (including ground corncobs) are readily available as bedding for laboratory rodents. Laminated-paper products are available for use in waste-collection pans, and shredded-paper products are marketed for use as contact bedding for rodents. Corncob and paper products treated with germicides or antibiotics to control bacterial growth are also available. However, the routine use of antibiotic-treated bedding materials might cause antibiotic-resistant strains of bacteria to develop or influence experimental results.

Bedding products manufactured specifically for use as rodent nesting materials are available. The use of such products, which might enhance neonatal survival in inbred rodent strains with inherently low reproduction rates, should be considered.

All rodent bedding products should be packaged in sealed, nonporous bags. Bags of bedding material should be stored in vermin-proof areas on pallets that do not touch the walls. When the bedding material is removed from the bags, it should be stored in metal or plastic containers that can be closed securely. The storage containers should be sanitized routinely.

SANITATION

Cleaning

Adequate sanitation is an integral part of maintaining laboratory rodents. Clean, sanitary conditions limit the presence of adventitious and opportunistic microorganisms, thereby decreasing their potential for compromising rodent health or causing adverse interactions with experimental procedures. Complete sterilization of the rodents' environment is seldom practical or necessary unless animals of highly defined microbiologic status or compromised immune status are used.

All components of the animal facility should undergo regular and thorough cleaning, including animal rooms, support areas (e.g., storage areas), cage-washing facilities, corridors, and procedure rooms. They should be cleaned with detergents and, when appropriate, disinfectant solutions to rid them of accumulated dirt and debris. Many such products are available. Selection of a cleaning agent should be based on how much and what kind of material is adhering to surfaces, as well as on the type of microbiologic contamination present (Block, 1991).

Monitoring of sanitation procedures should be appropriate to the process and materials used and might include visual inspection, monitoring of water temperatures, and microbiologic monitoring. It has been suggested that the effectiveness of sanitation procedures can be assessed by the intensity of animal odors, particularly ammonia; however, this should not be the sole means of assessing cleanliness, because too many variables are involved. Agents used to mask animal odors should not be used in rodent housing facilities; these agents cannot substitute for good sanitation practices, and their use exposes animals to volatile substances that can alter basic physiologic and metabolic processes.

The frequency with which surfaces are cleaned should be determined by how much use an area receives and the nature of potential contamination. Sweeping, mopping, and scrubbing with disinfectant agents should take place in a logical sequence. Cleaning utensils should be constructed of materials that resist corrosion and do not absorb dirt or debris. They should be stored in a neat, organized fashion. Wall-mounted hangers are useful for storing cleaning utensils because they reduce clutter, facilitate drying, and minimize contamination by keeping utensils off the floor. Cleaning utensils should be assigned to specific areas and should not be transported between areas. They should be regularly cleaned and dried, and there should be a regular schedule for replacing worn-out utensils.

Soiled bedding material should be removed and replaced with clean, dry bedding as often as is necessary to keep the animals clean and dry. The frequency is a matter of professional judgment and should be based on various factors, including the number and size of the animals housed in each cage, the anticipated urinary and fecal output, and the presence of debilitating conditions that might limit an animal's ability to access clean areas of the cage.

Bedding should be changed in a manner that reduces exposure of the animals and personnel to aerosolized waste materials. Laminar-flow bedding dump stations or similar devices can be used to control aerosol materials. If animals have been exposed to hazardous materials that are excreted in the urine or feces, additional precautions might be needed to prevent exposure of personnel while they are changing the bedding.

Frequent bedding changes can sometimes be counterproductive, for example, during portions of the postpartum period, changing the bedding removes pheromones, which are essential for successful reproduction (e.g., pheromones are necessary for synchronization of ovulation). Research objectives might also preclude frequent bedding changes. Under such circumstances, an exception to the regular bedding-change and cage-cleaning schedule can be justified.

Cages, cage racks, and accessory equipment, such as feeders and watering devices, should be cleaned and sanitized regularly to minimize the buildup of debris and to keep them free from contamination. Extra caging makes it easier to maintain a systematic schedule. Cleaning frequency will depend on the amount of bedding used, the frequency of bedding changes, the number of animals per cage, and other factors. In general, rodent cages and cage accessories will need to be washed at least once every 2 weeks. Solid-bottom rodent cages, water bottles, and sipper tubes usually require weekly cleaning. Some types of cage racking, large cages with very low animal density and frequent bedding changes, cages housing animals in gnotobiotic conditions, and cages used under other special circumstances might require less frequent cage-cleaning. Filter-top cages without forced-air ventilation and cages containing rodents with increased rates of production of feces or urine might require more frequent cleaning.

Cage-cleaning, debris removal, and disinfection can be accomplished in a single step or in multiple steps. Cage-cleaning and debris removal usually require the application of a detergent or surfactant solution coupled with mechanical action to remove adherent material from cage surfaces. Some laboratory rodents, such as guinea pigs and hamsters, produce urine with high concentrations of proteins and minerals. Their urine often binds aggressively to cage surfaces, which therefore require treatment with acid solutions before washing. Some detergents are rendered inactive at high temperatures, so, it is important to follow the manufacturer's instructions carefully.

Disinfection of cages is the process of killing vegetative forms of pathogenic bacteria. It can be accomplished by the action of either chemicals or hot water. If chemicals are used as the sole means of disinfection, careful attention should be paid to the concentration of the disinfectant solution's active ingredients, and the solution should be regularly changed in accordance with the manufacturer's instructions. When hot water is used either alone or in combination with disinfectant chemicals, temperatures and exposure times should be appropriate for adequate disinfection. Generally, the water temperature required for adequate disinfection precludes its use in anything but mechanical cage-washing equipment.

Cleaning and disinfection of cages can be done efficiently in mechanical cage washers. Washing times and conditions should be sufficient to kill vegetative forms of common bacteria and other microorganisms that are presumed to be controllable by sanitization. Microorganisms are killed by a combination of heat and the length of exposure to that heat (called the cumulative heat factor). Using high temperatures for short periods will produce the same cumulative heat factor and have the same effect on microorganisms as using lower temperatures for longer periods (Wardrip et al.,

1994). To achieve effective disinfection, water temperatures for washing and rinsing can vary from 58°C (143°F) to 82°C (180°F) or more. Recommendations for some types of mechanical cage washers using hot water alone for disinfection have been developed by the National Sanitation Foundation International (1990). Detergents and chemical disinfectants are known to enhance the effectiveness of hot water but must be thoroughly rinsed from surfaces to avoid harm to personnel and animals.

Cages and equipment can be effectively washed and disinfected by hand if appropriate attention is given to detail. Chemicals should be completely rinsed from surfaces, and personnel should have appropriate equipment to protect them from prolonged exposure.

Large pieces of caging equipment, such as racks, can be washed by hand; if large numbers are to be cleaned, portable cleaning equipment that dispenses detergent and hot water or steam under pressure might be more efficient. Large mechanical washing machines designed to accommodate racks and other pieces of large equipment are also commercially available.

Water bottles, sipper tubes, stoppers, and other small pieces of equipment should be washed with detergents, hot water, and, if appropriate, chemical agents to destroy vegetative forms of microorganisms. This process can be manual, if high-temperature rinse water is not used, or performed with mechanical washing equipment built especially for this purpose or a multiple-purpose cage-washing machine. Water bottles and sipper tubes can also be autoclaved after routine washing to ensure adequate sanitation.

If large numbers of water bottles or other small pieces of equipment are to be washed by hand, powered rotating brushes can be used to ensure adequate cleaning. Small items should be dipped or soaked in detergent and disinfectant solutions to maximize contact time. Therefore, large, two-compartment sinks are generally required if small items are to be hand washed.

If automatic watering systems are used, they should incorporate some mechanism to ensure that bacteria and debris do not build up in the watering devices. These systems are usually periodically flushed with large volumes of water or appropriate chemical agents and then rinsed to remove chemicals and associated debris. Constant-recirculation loops that use filters, ultraviolet light, or other treatment procedures to sterilize recirculated water can also be used.

Common methods of disinfection and sanitization are adequate for most rodent holding facilities. However, if pathogenic microorganisms are present or if rodents with highly defined microbiologic flora or compromised immune systems are maintained, it might be necessary to sterilize caging and other associated equipment after cleaning and disinfection. In such instances, access to an autoclave, gas sterilizer, or device capable of sterilizing with ionizing radiation is required. Whenever such sterilization processes are used, some form of regular monitoring is required to ensure their effectiveness.

Waste Containment and Disposal

Proper sanitation of an animal facility requires adequate containment, as well as regular and frequent removal of waste. Waste containers should be constructed of either metal or plastic materials and should be leakproof. They should be equipped with tight-fitting lids and, where appropriate, provided with disposable plastic liners for ease of waste removal. They should also be adequately labeled to distinguish between containers for hazardous and nonhazardous wastes; a color-coding system often proves useful.

If hazardous biologic waste is generated, an inventory sheet might be necessary for each waste container, so that the type of waste and the approximate quantity of hazardous material can be recorded. Waste containers for animal tissues or carcasses should be lined with leakproof, disposable liners that will withstand being refrigerated or frozen to reduce tissue decomposition. If wastes are collected and stored before removal from the site, the storage area should be physically separated from other facilities used to house animals or store animal-related materials. Waste-storage areas should be cleaned regularly and kept free of insects and other vermin. All waste containers and associated implements should be cleaned and disinfected frequently.

Waste materials from rodent housing facilities can be disposed of in various ways (depending on the type of waste), including incineration, agricultural composting, and landfill disposal. Hazardous waste must be separated from other waste, and its classification and handling are controlled by a variety of local, state, and federal agencies. Some form of pretreatment—such as autoclaving, chemical neutralization, or compaction with absorbents—might be required. The National Safety Council (1979) has recommended procedures for disposal of hazardous waste. It is the institution's responsibility to comply with all federal, state, and municipal statutes and ordinances regarding the control, movement, and disposal of hazardous waste.

Pest Control

All rodent housing facilities should have a program to prevent, control, or eliminate infestation by pests (including insects and wild and escaped rodents). The program should include regular inspection of the premises for signs of pests, a monitoring system that uses rodent traps and insect-collection devices to capture pests, and regular evaluation of the integrity and condition of the animal facilities. The pest-control program should focus on preventing the entry of vermin into the facility (by sealing potential points of entry and eliminating sites outside the facility where vermin can breed or be harbored) and maintaining an environment in which pests cannot sustain themselves and reproduce. Only if those methods are ineffective should the use of pesticides be considered.

If pesticides are required, relatively nontoxic substances (e.g., boric acid, amorphous silica gel, and insect-growth regulating hormones) and mechanical devices (e.g., adhesive traps, air curtains, and insect-electrocution devices) should be used in preference to toxic materials, especially for controlling insect pests. If a toxic compound is to be used in animal areas, it should be used only after consultation with the investigators whose animals are housed in the facility because of potential effects on the animals' health and possible interference with research results. The application of toxic pesticides should be coordinated with those responsible for the management of the animal-care program and carried out by licensed applicators in compliance with local, state, and federal regulations.

The pest-control program should be adequately documented, including records of dates and methods of application of pesticides and possibly records of inspection, results of monitoring and trapping programs, records of sightings and identification of pests, and maintenance schedules.

IDENTIFICATION AND RECORDS

Adequate individual or group identification of rodents and appropriate records of their care and use are essential to the conduct of biomedical research programs. Individual identification of rodents is not always required; when necessary, it can be accomplished in various ways, including ear-punching, use of ear tags, tattooing (usually on the tail), or implanting electromagnetic transponders. If ear tags are used, they should be light enough so that they do not visibly change the animal's head posture, and surrounding tissues should be monitored for inflammation. Dyes are occasionally used on the fur, skin, or tail for temporary identification. In general, amputation of digits (toe-clipping) is no longer an acceptable method of identification, because more humane methods can usually be substituted.

Individual animals or groups of animals can also be identified with cage identification cards. If cards are used, sufficient information is required to identify and characterize the animals in the cage adequately. This information can include such details as the name and location (e.g., office location, telephone number, and division or department name) of the responsible investigator; the species, strain, or stock of the animals; the sex of the animals; the number of animals in the cage; the source of the animals; institutional identification numbers (e.g., IACUC-approved protocol number and purchase-order number); and, when appropriate, other identifying information pertaining to the project (e.g., group designation and age or weight specifications). Bar-code identifiers can also be included on the cage card to aid in identifying the animals and linking their identification with other, more detailed records. Color-coding the cage cards and labeling cage racks and animal holding rooms are effective management tools for locating and identifying animals.

Some research protocols require that records be kept on individual animals, for example, when animals are used in breeding programs or are exposed to hazardous agents. Detailed surgical records are not commonly maintained on individual rodents but might be helpful in some situations such as when complex surgical procedures are being used or when new procedures are being developed.

RODENTS OTHER THAN RATS AND MICE

Guinea Pigs

One of the most striking ways in which guinea pigs (*Cavia porcellus*) differ from rats and mice is the guinea pigs' absolute requirement for exogenous vitamin C, a requirement that is shared with humans and only a few other species. Because of that requirement, guinea pig diets must be fortified with vitamin C. As an alternative, vitamin C can be added to the drinking water or provided in the form of food supplements, including such vegetables as kale, that are high in vitamin C. The use of food supplements should be approached with some caution because of the possibility of contamination with chemicals or microorganisms that could influence the course of experimentation. Vitamin C is a very labile compound, so storage conditions of foods containing it and heat treatment of such foods, including autoclaving, are of particular concern.

The guinea pigs' body conformation makes design and placement of feeders important. Feeders should be designed to avoid trauma to the chin and neck area of guinea pigs. Guinea pigs will occasionally rear up on their hind legs, but they will not accept food from feeders suspended overhead. Bowls for food and water can be used instead of more conventional feeding and watering devices; but guinea pigs like to nest in such receptacles, and that causes waste and contamination of food. Feeders that have a J shape are best suited to address these concerns and are used most commonly.

Guinea pigs, like other rodents, tend to eat and drink throughout the day and night. They become habituated to a particular diet and have defined taste preferences. Any changes in the composition of the food—especially changes in size, shape, consistency, or taste—can cause a sharp decline in food consumption. If the animals fail to adapt to the new food, severe weight loss or even starvation and death can occur; therefore, new food should be introduced gradually.

Guinea pigs often grow to weigh more than 1 kg and have relatively small feet. They have a well-developed startle response that causes them to make sudden movements in response to unfamiliar sounds; when they are housed in groups, this might be manifested as a stampede. Those two traits make cage-floor design particularly important. Wire-bottom cages should be designed to provide sufficient support for the animals' feet to prevent pressure sores, and the space between the wires in the floor grid should be small enough to preclude entrapment of animals' feet.

Guinea pigs also differ substantially from rats and mice in having a vaginal closure membrane and a long gestation period. Gestation in guinea pigs can range from 59 to 72 days; 63 to 68 days is the average. Gestation length can be affected by several characteristics, including litter size, which is usually one to three pups (McKeown and Macmahon, 1956). Female and male guinea pigs reach puberty as early as 4-5 weeks old and 8-10 weeks old, respectively, but are best mated when 2.5-3 months old or when they weigh 450-600 g (Ediger, 1976). Because a relatively large fetal mass is expelled at parturition, a female should be bred before she is 6 months old to minimize the likelihood of being excessively fat or having firm fusion of the symphysis pubis. If the symphysis pubis is fused, it cannot separate the approximate 0.5 in. needed for passage of fetuses through the birth canal; the result can be severe reproductive problems and death of both fetus and mother.

Strain 13 guinea pigs, which are highly inbred, should be housed to protect them from or immunized against the common bacterium *Bordetella bronchiseptica* (Ganaway et al., 1965). Treating guinea pigs for bacterial infections should be approached with caution because antibiotics can cause acute effects. Some can be administered safely; others, such as penicillin, can cause toxemia and death (Pakes et al., 1984; Wagner, 1976). The problem appears to be associated with the excretion of the antibiotics into the gastrointestinal tract and the resulting disturbance of the microbiologic flora on which the guinea pig depends for much of its digestive processes.

Guinea pigs produce large volumes of urine that contain substantial quantities of dissolved minerals and protein. Their urine adheres tenaciously to surfaces, and soaking in dilute solutions of organic acids is often required before cages are cleaned. Urination and dragging the perineum across the floor of the cage are common methods by which guinea pigs mark freshly cleaned cages.

Hamsters

Laboratory hamsters belong to the subfamily Cricetidae. The most common and most readily available commercially is the Syrian hamster, *Mesocriçetus auratus* (sometimes called the golden hamster). Syrian hamsters are native to arid regions of the Middle East and have become well adapted to conserving water, which they obtain principally through food. In a laboratory environment, hamsters will drink water from water bottles, bowls, or automatic watering systems. Hamsters secrete highly concentrated urine that contains large quantities of mineral salts; their urine tends to leave deposits on cage surfaces that are often difficult to remove and might require the application of dilute acids.

Hamsters are often aggressive toward each other, and care should be taken when they are housed in groups. Hamsters that fight must be separated to prevent injury. Cannibalization can occur in group-housed animals when an animal becomes sick or debilitated. It is important to separate animals that are observed to be clinically abnormal.

Vitamin E is an important nutritional requirement of hamsters; vitamin E deficiency has been associated with muscular dystrophy (West and Mason, 1958) and fetal central nervous system hemorrhagic necrosis (Keeler and Young, 1979). Most commercial rodent diets are supplemented with vitamin E, but care is required to ensure the adequacy of vitamin E if special-formula, purified, or semipurified diets are used (Balk and Slater, 1987). The method of food presentation is important. If food is placed in suspended feeders, hamsters will remove it from the feeder and pile it on the floor. Location of the food pile is peculiar to individual hamsters and will vary from one cage environment to the next. Moving food away from a pile will cause the hamsters to retrieve it and move it back. Given that behavioral pattern, feeding hamsters on the floor of the cage is considered acceptable (9 CFR 3.29). Hamsters have cheek pouches in which they hold and transport food; a full cheek pouch should not be mistaken for a pathologic condition.

Hamsters have very loose skin, particularly over the shoulders. Care should be taken when picking them up so that they do not turn around and bite the handler. Hamsters can be tamed by regular, gentle handling. Without such taming, they can be aggressive toward the handler.

Many species of hamsters hibernate if conditions are right. Various environmental influences seem important, including seasonality, photoperiod, ambient temperature, availability of food, and isolation. To avoid hibernation, temperatures should be maintained within ranges specified in the *Guide* (NRC, 1996 et seq.).

Hamsters, like guinea pigs, are susceptible to antibiotic associated toxicity and enterocolitis. Although successful use of antibiotics in hamsters has been reported, the reports usually involve smaller than therapeutic dosages of antibiotics or the use of particular antibiotic preparations that are not excreted into the gastrointestinal tract (Pakes et al., 1984; Small, 1987). As a general rule, antibiotics should be avoided in hamsters.

Estrus in hamsters is similar to that in mice, lasting 4-5 days; however, the gestation period is considerably shorter than that in mice—an average of 16 days. Hamsters are commonly pair-mated; the female is taken to the male's cage for breeding on detection of a stringy vaginal discharge that occurs when the female is in estrus. The female can be removed from the male's cage after mating is observed; however, conception is sometimes improved by leaving her with the male for 24 hours. Removing the female after that time minimizes fighting and allows the male to breed with other females. For optimal reproduction, the light cycle should be maintained at 14 hours of light and 10 hours of dark, which is slightly different from that for other rodents. Litter size ranges from 4 to 16 pups; first litters tend to be smaller than subsequent litters. Cannibalism of pups is common, especially in first litters. It is important to furnish enough bedding or nesting material for the neonates to stay well hidden and to provide the dam with enough food to allow her to be undisturbed from about 2-3 days before birth until about 7-10 days after birth (Balk and Slater, 1987; Harkness and Wagner, 1989).

Gerbils

Gerbils (*Meriones unguiculatus*) do well in solid-bottom cages. Gerbils tend to stand and sit upright and often exhibit a digging or scratching behavior in the corners of cages while in an upright posture. Therefore, cages that are tall enough for this behavior are generally preferred.

Gerbils tend to form social relationships early in life, and groups established at puberty tend to exhibit minimal fighting or other aggressive behavior; aggressive behavior is more common when individual animals are put together later in life. New mates are not accepted easily. For those reasons, it is prudent to select a paired-mating scheme for establishment of colonies and not to regroup gerbils often.

Estrus in gerbils lasts 4-6 days; gestation in nonlactating females is about 24-26 days. If females are bred in the postpartum period, implantation is delayed, and gestation can be as long as 48 days. To avoid postpartum mating, the male can be removed from the cage, but he should be returned to his mate within 2 weeks to decrease the possibility of fighting (Harkness and Wagner, 1989). Average litter size is 3-7.

Gerbils are generally very tame and rarely bite unless mishandled. When they are excited, they will jump and dart about to resist being caught. Gerbils should not be suspended by holding their tails, because the skin over the tail is relatively loose and can be pulled off easily.

Commercial rodent diets are usually acceptable for gerbils, provided that they have a low fat content. Because of the gerbils' unique fat metabolism, it is not uncommon for them to develop high blood cholesterol concentrations on diets containing fat at 4 percent or more. When fed a diet high in fat, gerbils tend to store the fat and become obese. In females, the fat accumulation can be associated with reproductive difficulty.

Chinchillas

Chinchillas (*Chinchilla laniger*) have been farmed for pelts since 13 animals were imported from South America to California in 1927. Most domestic stock is believed to be descended from those animals (Anderson and Jones, 1984). Chinchillas can be housed in wire-mesh or solid-bottom cages; the latter are preferred for breeding (Clark, 1984; Weir, 1976). They are fastidious groomers and should be provided with a box containing a mixture of silver sand and Fuller's earth for a short period daily to allow dust bathing (Clark, 1984). Chinchillas tolerate cold but are very sensitive to heat; the suggested temperature is 20°C (68°F) (Weir, 1976). Commercial chinchilla feed is available, but standard guinea pig rations can also be used (Clark, 1984; Weir, 1976). They might require a source of roughage, such as hay (Weir, 1967). Water and food should be made available ad libitum.

The system used most commonly for breeding chinchillas is to put one male with several females in a large cage. However, females are larger than males and are very aggressive toward both males and other females, and it is necessary to provide refuges, such as nesting boxes, for animals that are being attacked. An "Elizabethan collar" can be used to keep an aggressive female from following an animal that she is attacking into its refuge. A light:dark ratio of 14:10 hours is adequate (Weir, 1967). The mean gestation period is 111 days, with a range of 105-118 days (Clark, 1984). Chinchilla litter size ranges from one to six, with a mean of two. The young are born fully furred and with open eyes, and they begin eating solid food within 1 week but are not completely weaned until they are 6-8 weeks old. Females do not build nests.

REFERENCES

Algers, B., I. Ekesbo, and S. Stromberg. 1978. The impact of continuous noise on animal health. Acta Vet. Scand. 67(Suppl.):1-26.

Alleva, J. J., M. V. Waleski, F. R. Alleva, and E. J. Umberger. 1968. Synchronizing effect of photoperiodicity on ovulation in hamsters. Endocrinology 82:1227-1235.

Anderson, K. V., F. P. Coyle, and W. K. O'Steen. 1972. Retinal degeneration produced by low-intensity colored light. Exp. Neurol. 35:233-238.

Anderson, S., and J. K. Jones, Jr., eds. 1984. Orders and Families of Recent Mammals of the World. New York: John Wiley and Sons. 686 pp.

Anthony, A., and J. E. Harclerode. 1959. Noise stress in laboratory rodents. II: Effects of chronic noise exposures on sexual performance and reproductive function of guinea pigs. J. Acoust. Soc. Am. 31:1437-1440.

ASHRAE (American Society Heating, Refrigeration, and A Engineers, Inc.). 1993. Chapter 9: Environmental Control for Animals and Plants. In 1993 ASHRAE Handbook: Fundamentals, I-P edition. Atlanta: ASHRAE

Baetjer, A. M. 1968. Role of environmental temperature and humidity in susceptibility to disease. Arch. Environ. Health 16:565-570.

Balk, M. W., and G. M. Slater. 1987. Care and management. Pp. 61-67 in Laboratory Hamsters, G. L. Van Hoosier, Jr., and C. W. McPherson, eds. Orlando, Fla.: Academic Press.

Bank, H. L., J. John, M. K. Schmehl, and R. J. Dratch. 1990. Bactercidal effectiveness of modulated UV light. Appl. Environ. Microbiol. 56:3888-3889.

Barkley, W. E. 1978. Abilities and limitations of architectural and engineering features in controlling biohazards in animal facilities. Pp. 158-163 in Laboratory Animal Housing. Proceedings of a symposium organized by the ILAR Committee on Laboratory Animal Housing and held September 22-23, 1976, in Hunt Valley, Maryland. Washington, D.C.: National Academy of Sciences.

Barnett, S. A. 1955. Competition among wild rats. Nature 175:126-127.

Barrett, A. M., and M. A. Stockham. 1963. The effect of housing conditions and simple experimental procedures upon the corticosterone level in the plasma of rats. J. Endocrinol. 26:97-105.

Bell, R. W., C. E. Miller, J. M. Ordy, and C. Rolsten. 1971. Effects of population density and living space upon neuroanatomy, neurochemistry, and behavior in the C57B1-10 mouse. J. Comp. Physiol. Psychol. 75:258-263.

Bellhorn, R. W. 1980. Lighting in the animal environment. Lab. Anim. Sci. 30:440-450.

Besch, E. L. 1975. Animal cage from dry bulb and dewpoint temperature differentials. ASHRAE Trans. 81:549-558.

Besch, E. L. 1980. Environmental quality within animal facilities. Lab. Anim. Sci. 30:385-406.

Besch, E. L. 1985. Definition of laboratory animal environmental conditions. Pp. 297-315 in Animal Stress, G. P. Moberg, ed. Bethesda, Md.: American Physiological Society.

Blackmore, D. 1970. Individual differences in critical temperatures among rats at various ages. J. Appl. Physiol. 29:556-559.

Block, S. S., ed. 1991. Disinfection, Sterilization, and Preservation. 4th ed. Philadelphia: Lea & Febiger. 1,162 pp.

Bock, G. R., and J. C. Saunders. 1977. A critical period for acoustic trauma in the hamster and its relation to cochlear development. Science 197:396-398.

Brain, P., and D. Benton. 1979. The interpretation of physiological correlates of differential housing in laboratory rats. Life Sci. 24:99-115.

Brainard, G. C. 1988. Illumination of animal quarters in microgravity habitats: Participation of light irradiance and wavelength in the photo regulation of the neuroendocrine system. Pp. 217-252 in Lighting Requirements in Microgravity—Rodents and Nonhuman Primates, D. C. Holley, C. M. Winget, and H. A. Leon, eds. NASA Technical Memorandum 101077. Washington, D.C.: National Aeronautics and Space Administration.

Brainard, G. C. 1989. Illumination of laboratory animal quarters: Participation of light irradiance and wavelength in the requlation of the neuroendocrine system. Pp. 69-74 in Science and Animals: Addressing Contemporary Issues, H. N. Guttman, J. A. Mench, and R. C. Simmonds, eds. Bethesda, Md.: Scientists Center for Animal Welfare. Available from SCAW, Golden Triangle Building One, 7833 Walker Drive, Suite 340, Greenbelt, MD 20770.

Broderson, J. R., J. Lindsey, and J. E. Crawford. 1976. The role of environmental ammonia in respiratory mycoplasmosis of rats. Am. J. Pathol. 85:115-130.

Bucci, T. J. 1992. Dietary restriction: Why all the Interest? An overview. Lab Anim. 21(6):29-34.

Burdick, C. K., J. H. Patterson, and B. T. Mozo, R.T. Camp, Jr.. 1978. Threshold shifts in chinchillas exposed to octave bands of noise centered at 63 and 1000 Hz for three days (a). J. Acoust. Soc. Am. 64:458-466.

CCAC (Canadian Council on Animal Care). 1980. Guide to the Care and Use of Experimental Animals, Vol. 1. Ottawa: Canadian Council on Animal Care. 120 pp. Available from CCAC, Constitution Square, Tower II, 315-350 Albert, Ottawa, Ontario, Canada K1R 1B1.

Christian, J. J. 1960. Adrenocortical and gonadal responses of female mice to increased population density. Proc. Soc. Exp. Biol. Med. 104:330-332.

Christian, J. J., and C. D. LeMunyan. 1958. Adverse effects of crowding on lactation and reproduction of mice and two generations of their progeny. Endocrinology 63:517-529.

Clark, J. D. 1984. Biology and diseases of other rodents. Pp. 183-205 in Laboratory Animal Medicine, J. G. Fox, B. J. Cohen, and F. M. Loew, eds. Orlando, Fla.: Academic Press.

Clarke, H. E., M. E. Coates, J. K. Eva, D. J. Ford, C. K. Milner, P. N. O'Donoghue, P. P. Scott, and R. J. Ward. 1977. Dietary standards for laboratory animals: Report of the Laboratory Animals Centre Diets Advisory Committee. Lab. Anim. (London) 11:1-28.

Clough, G. 1976. The immediate environment of the laboratory animal. Pp. 77-94 in Control of the Animal House Environment, T. McSheehy, ed. Laboratory Animal Handbooks 7. London: Laboratory Animals Ltd.

Coates, M. E., ed. 1987. ICLAS Guidelines on the Selection and Formulation of Diets for Animals in Biomedical Research. London: Institute of Biology.

Coates, M. E., J. E. Ford, M. E. Gregory, and S. Y. Thompson. 1969. Effects of gamma-irradiation on the vitamin content of diets for laboratory animals. Lab. Anim. (London) 3:39-49.

Council of Europe. 1990. European Convention for the Protection of Vertebrate Animals Used for Experimental and Other Scientific Purposes. Strasbourg: Council of Europe. 53 pp.

Cunliffe-Beamer, T. L., L. C. Freeman, and D. D. Myers. 1981. Barbiturate sleeptime in mice exposed to autoclaved or unautoclaved wood beddings. Lab. Anim. Sci. 31:672-675.

Curd, E. F. 1976. Heat losses and heat gains. Pp. 153-183 in Control of the Animal House Environment, T. McSheehy, ed. Laboratory Animal Handbooks 7. London: Laboratory Animals Ltd.

Davis, D. E. 1958. The role of density in aggressive behavior of house mice. Anim. Behav. 6:207-210.

Davis, T. A., C. W. Bales, and R. E. Beauchene. 1983. Differential effects of dietary caloric and protein restriction in the aging rat. Exp. Gerontol. 18:427-435.

Dunklin, E. W., and T. T. Puck. 1948. The lethal effect of relative humidity on airborne bacteria. J. Exp. Med. 87:87-101.

Dyment, J. 1976. Air filtration. Pp. 209-246 in Control of the Animal House Environment, T. McSheehy, ed. Laboratory Animal Handbooks 7. London: Laboratory Animals Ltd.

Ediger, R. D. 1976. Care and management. Pp. 5-12 in The Biology of the Guinea Pig, J. E. Wagner and P. J. Manning, eds. New York: Academic Press.

Engelbrecht, R. S., M. J. Weber, B. L. Salter, and C. A. Schmidt. 1980. Comparative inactivation of viruses by chlorine. Appl. Environ. Microbiol. 40:249-256.

Ferguson, H. C. 1966. Effect of red cedar chip bedding on hexobarbital and phenobarbital sleep time. J. Pharm. Sci. 55:1142-1143.

Fidler, I. J. 1977. Depression of macrophages in mice drinking hyperchlorinated water. Nature 270:735-736.

Flynn, R. J. 1959. Studies on the aetiology of ringtail of rats. Proc. Anim. Care Panel 9:155-160.

Flynn, R. J. 1968. A new cage cover as an aid to laboratory rodent disease control. Proc. Soc. Exp. Biol. Med. 129:714-717.

Foster, H. L., C. L. Black, and E. S. Pfau. 1964. A pasteurization process for pelleted diets. Lab. Anim. Care 14:373-381.

Fox, J. G., F. D. Aldrich, and G. W. Boylen, Jr. 1976. Lead in animal foods. J. Toxicol. Environ. Health 1:461-467.

Gamble, M. R., and G. Clough. 1976. Ammonia build-up in animal boxes and its effect on rat tracheal epithelium. Lab. Anim. (London) 10(2):93-104.

Ganaway, J. R., A. M. Allen, and C. W. McPherson. 1965. Prevention of acute *Bordetella bronchiseptica* pneumonia in a guinea pig colony. Lab. Anim. Care 15:156-162.

Geber, W. F. 1973. Inhibition of fetal osteogenesis by maternal noise stress. Fed. Proc. 32:2101-2104.

Geber, W. F., T. A. Anderson, and B. Van Dyne. 1966. Physiologic responses of the albino rat to chronic noise stress. Arch. Environ. Health 12:751-754.

Goodrick, C. L. 1978. Body weight increment and length of life: The effect of genetic constitution and dietary proteins. J. Gerontol. 33:184-190.

Green, D. E., and P. K. Stumpf. 1946. The mode of action of chlorine. J. Amer. Water Works Assoc. 38:1301-1305.

Green, G. H. 1974. The effect of indoor relative humidity on absenteeism and colds in schools. ASHRAE Trans. 80(2):131-141.

Greenman, D. L., P. Bryant, R. L. Kodell, and W. Sheldon. 1982. Influence of cage shelf level on retinal atrophy in mice. Lab. Anim. Sci. 32:353-356.

Guha, D., E. F. Williams, Y. Nimitkitpaisan, S. Bose, S. N. Dutta, and S. N. Pradhar. 1976. Effects of sound stimulus on gastric secretion and plasma corticosterone level in rats. Res. Commun. Chem. Pathol. Pharmacol. 13:273-281.

Hall, J. E., W. J. White, and C. M. Lang. 1980. Acidification of drinking water: Its effects on seleccted biologic phenomena in male mice. Lab. Anim. Sci. 30:643-651.

Hann, V. 1965. Disinfection of drinking water with ozone. J. Am. Water Works Assoc. 48:1316.

Harkness, J. E., and J. E. Wagner. 1989. Biology and husbandry. Pp. 9-54 in The Biology and Medicine of Rabbits and Rodents, 3rd ed. Philadelphia: Lea & Febiger.

Harstad, J. B., H. M. Decker, L. M. Buchanan, and M. E. Filler. 1967. Air filtration of submicron virus aerosols. Am. J. Public Health Nations Health 57:2186-2193.

Helrich, K. ed. 1990. Official Methods of Analysis of the Association of Official Analytical Chemists, 15th ed. Arlington, Va.: Association of Official Analytical Chemists (AOAC). Available from AOAC, 2200 Wilson Boulevard, Suite 400, Arlington, VA 22109-3301.

Hermann, L. M., W. J. White, and C. M. Lang. 1982. Prolonged exposure to acid, chlorine, or tetracycline in drinking water: effects on delayed-type hypersensitivity, hemagglutination titers, and reticuloendothelial clearance rates in mice. Lab. Anim. Sci. 32:603-608.

Holick, M. F. 1989. Cutaneous synthesis of vitamin D: Can dietary vitamin D supplemetation substitute for sunlight? Pp. 63-68 in Science and Animals: Addressing Contemporary Issues, H. N. Guttman, J. A. Mench, and R. C. Simmonds, eds. Bethesda, Md.: Scientists Center for Animal Welfare. Available from SCAW, Golden Triangle Building One, 7833 Walker Drive, Suite 340, Greenbelt, MD 20770.

Hughes, P. C., and M. Nowak. 1973. The effect of the number of animals per cage on the growth of the rat. Lab. Anim. (London) 7:293-296.

Iwasaki, K., C. A. Gleiser, E. J. Masoro, C. A. McMahan, E.-J. Seo, and B. P. Yu. 1988. Influence of the restriction of individual dietary components on longevity and age-related disease of Fischer rats: the fat component and the mineral component. J. Gerontol. 43:B13-B21.

Joasoo, A., and J. M. McKenzie. 1976. Stress and the immune response in rats. Int. Arch. Allergy Appl. Immunol. 50:659-663.

Johnstone, M. W., and P. F. Scholes. 1976. Measuring the environment. Pp. 113-128 in Control of the Animal House Environment, T. McSheehy, ed. Laboratory Animal Handbooks 7. London: Laboratory Animals Ltd.

Kaufman, J. E., ed. 1987. IES Lighting Handbook. New York: Illuminating Engineering Society of North America.

Keeler, R. F., and S. Young. 1979. Role of vitamin E in the etiology of spontaneous hemorrhagic necrosis of the central nervous system of fetal hamsters. Teratology 20:127-32.

Keenan, K. P., P. F. Smith, and K. A. Soper. 1994. Effect of dietary (caloric) restriction on aging, survival, pathology, and toxicology. Pp. 609-628 in Pathobiology of the Aging

Rat, vol. 2, W. Notter, D. L. Dungworth, and C. C. Capen, eds. International Life Sciences Institute.

Keller, L. S., W. J. White, M. T. Snyder, and C. M. Lang. 1989. An evaluation of intra-cage ventilation in three animal caging systems. Lab. Anim. Sci. 39:237-242.

Kelly, J. B., and B. Masterton. 1977. Auditory sensitivity of the albino rat. J. Comp. Physiol. Psychol. 91:930-936.

Kimmel, C. A., R. O. Cook, and R. E. Staples. 1976. Teratogenic potential of noise in mice and rats. Toxicol. Appl. Pharmacol. 36:239-245.

Knapka, J. J. 1983. Nutrition. Pp. 51-67 in The Mouse in Biomedical Research. Vol. III: Normative Biology, Immunology, and Husbandry, H. L. Foster, J. D. Small, and J. G. Fox, eds. New York: Academic Press.

Knapka, J. J. 1985. Formulation of diets. Pp. 45-59 in Methods for Nutritional Assessment of Fats, J. Beare-Rogers, ed. Champaign, Ill.: American Oil Chemists Society. Available from the American Oil Chemists Society, PO Box 3489, Champaign, IL 61826-3489.

Knapka, J. J., K. P. Smith, and F. J. Judge. 1974. Effect of open and closed formula rations on the performance of three strains of laboratory mice. Lab. Anim. Sci. 24:480-487.

Kool, H. J., and J. Hrubec. 1986. The influence of ozone, chlorine and chlorine dioxide treatment on mutagenic activity in drinking water. Ozone Sci. Eng. 8(3):217.

Kraak, W., and G. Hofmann. 1977. Detection of noise-induced physiological stress and hearing loss in guinea pigs by means of an electrochleographic method. Arch. Otorhinolaryngol. 215:301-310.

Kubo, C., B. C. Johnson, N. K. Day, and R. A. Good. 1984. Calorie source, caloric restriction, immunity, and aging of (NZB/NZW) F^1 mice. J. Nutr. 114:1884-1899.

Lai, Y.-L., R. O. Jacoby, and A. M. Jonas. 1978. Age-related and light-associated retinal changes in Fischer rats. Invest. Ophthalmol. Vis. Sci. 17:634-638.

LaVail, M. M. 1976. Rod outer segment disk shedding in rat retina: relationship to cyclic lighting. Science 194:1071-1074.

Lawlor, M. 1990. The size of rodent cages. Pp. 19-28 in Guidelines for the Well-being of Rodents in Research, H. N. Guttman, ed. Proceedings from a conference organized by the Scientists Center for Animal Welfare and held December 8, 1989, in Research Triangle Park, North Carolina. Bethesda, Md.: Scientists Center for Animal Welfare.

Lee, R. C. 1942. Heat production of the rabbit at 28°C as affected by previous adaptation to temperature between 10° and 31°C. J. Nutr. 23(1):83-90.

Ley, F. J., J. Bleby, M. E. Coates, and J. S. Patterson. 1969. Sterilization of laboratory animal diets using gamma radiation. Lab. Anim. (London) 3:221-254.

Lipman, N. S., B. F. Corning, and M. A. Coiro. 1992. The effects of intracage ventilation on microenvironmental conditions in filter-top cages. Lab. Anim. (London) 26:206-210.

McEllhiney, R., ed. 1985. Feed Manufacturing Technology III. Arlington, Va.: American Feed Industry Association. 602 pp. Available from the American Feed Industry Association, 1501 Wilson Boulevard, Arlington, VA 22209.

McKeown, T., and B. Macmahon. 1956. The influence of litter size and litter order on length of gestation and early postnatal growth in guinea pigs. Endocrinology 13:195-200.

Meier, H., and M. C. Hoag. 1966. Blood coagulation. Pp. 373-376 in Biology of the Laboratory Mouse, 2d ed., E. L. Green, ed. New York: McGraw-Hill Book Co.

Mills, C. A. 1945. Influence of environmental temperatures on warm-blooded animals. Ann. N.Y. Acad Sci. 46(1):97-105.

Mills, C. A., and L. H. Schmidt. 1942. Environmental temperatures and resistance to infection. Am. J. Trop. Med. 22:655-660.

Moller, A. 1978. Review of animal experiments. J. Sound Vibr. 59:73-77.

Munkelt, H. F. 1938. Odor control in animal laboratories. Heat. Piping Air Cond. 10:289-291.

Murakami, H. 1971. Differences between internal and external environments of the mouse cage. Lab. Anim. Sci. 21(5):680-684.

National Safety Council. 1979. Disposal of Potentially Contaminated Animal Wastes. Data Sheet 1-167-79. Chicago: National Safety Council.

National Sanitation Foundation International. 1990. Standard 3: Commercial Spray-type Dishwashing Machines. Ann Arbor, Mich.: National Sanitation Foundation International. Available from the National Sanitation Foundation International, 3475 Plymouth Road, PO Box, 130140, Ann Arbor, MI 48113-0140 (telephone, 313-769-8010).

Navia, J. M. 1977. Preparation of diets used in dental research. Pp. 151-167 in Animal Models in Dental Research. University, Ala.: University of Alabama Press.

Nayfield, K. C., and E. L. Besch. 1981. Comparative responses of rabbits and rats to elevated noise. Lab. Anim. Sci. 31:386-390.

Nevins, R. G., and P. L. Miller. 1972. Analysis, evaluation and comparison of room air distribution performance—A summary. ASHRAE Trans. 28(2):235-242.

Newberne, P. M. 1975. Influence on pharmacological experiments of chemicals and other factors in diets of laboratory animals. Fed. Proc. 34:209-218.

Newell, G. W. 1980. The quality, treatment, and monitoring of water for laboratory rodents. Lab. Anim. Sci. 30(2, part II):377-384.

Njaa, L. R., F. Utne, and O. R. Braekkan. 1957. Effect of relative humidity on rat breeding and ringtail. Nature 180:290-291.

NRC (National Research Council), Institute of Laboratory Animal Resources, Committee on Care and Use of Laboratory Animals. 1978. Guide for the Care and Use of Laboratory Animals. DHEW Pub. No. (NIH) 78-23. Washington, D.C.: U.S. Department of Health, Education, and Welfare. 70 pp.

NRC (National Research Council), Institute of Laboratory Animal Resources, Committee to Revise the Guide for the Care and Use of Laboratory Animals. 1996. Guide for the Care and Use of Laboratory Animals, 7th edition. Washington, D.C.: National Academy Press.

NRC (National Research Council), Board on Agriculture, Committee on Animal Nutrition, Subcommittee on Laboratory Animal Nutrition. 1995. Nutrient Requirements of Laboratory Animals, 4th revised ed. Nutrient Requirements of Domestic Animals Series. Washington, D.C.: National Academy Press.

Ogle, C. 1934. Climatic influence on the growth of the male albino mouse. Am. J. Physiol. 107:635-640.

O'Steen, W. K. 1970. Retinal and optic nerve serotonin and retinal degeneration as influenced by photoperiod. Exp. Neurol. 27:194-205.

Pakes, S. P., Y.-S. Yu, and P. C. Meunier. 1984. Factors that complicate animal research. Pp. 649-665 in Laboratory Animal Medicine, J. G. Fox, B. J. Cohen, and F. M. Loew, eds. Orlando, Fla.: Academic Press.

Peterson, E. A. 1980. Noise and laboratory animals. Lab. Anim. Sci. 30:2 Pt. II 422-439.

Peterson, E. A., J. S. Augenstein, D. C., Tanis, and D. G. Augenstein. 1981. Noise raises blood pressure without impairing auditory sensitivity. Science 211:1450-1452.

Pleasants, J. R. 1984. Diets for germ-free animals. Part 2: The germ-free animal fed chemically defined ultrafiltered diet. Pp. 91-109 in The Germ-Free Animal in Biomedical Research, M. E. Coates and B. E. Gustafsson, eds. London: Laboratory Animals Ltd.

Pleasants, J. R., M. H. Johnson, and B. S. Wostmann. 1986. Adequacy of chemically defined, water-soluble diet for germ free BALB/c mice through successive generations and litters. J. Nutr. 116:1949-1964.

Poole, T. B., and H. D. R. Morgan. 1976. Social and territorial behavior of laboratory mice (*Mus musculus* L.) in small complex areas. Anim. Behav. 24:476-480.

Porter, G., and W. Lane-Petter. 1965. The provision of sterile bedding and nesting materials with their effects on breeding mice. J. Anim. Tech. Assoc. 16:5-8.

Rao, G. N. 1990. Long-term toxicological studies using rodents. Pp. 47-52 in Guidelines for the Well-being of Rodents in Research, H. N. Guttman, ed. Proceedings from a conference organized by the Scientists Center for Animal Welfare and held December 8, 1989, in Research Triangle Park, North Carolina. Bethesda, Md.: Scientists Center for Animal Welfare.

Rao, G. N., and J. J. Knapka. 1987. Contaminant and nutrient concentrations of natural ingredient rat and mouse diet used in chemical toxicology studies. Fundam. Appl. Toxicol. 9:329-338.

Reiter, R. J. 1991. Pineal gland: Interface between the photoperiodic environment and the endocrine system. Trends Endocrinol. Metab. 2:13-19.

Remé, C. E., A. Wirz-Justice, and M. Terman. 1991. The visual input stage of the mammalian circadian pacemaking system. I. Is there a clock in the mammalian eye?. J. Biol. Rhythms 6(1):5-29.

Runkle, R. S. 1964. Laboratory animal housing—Part II. J. Am. Inst. Arch. 41:77-80.

Scharmann, W. 1991. Improved housing of mice, rats and guinea pigs: A contribution to the refinement of animal experiments. ATLA 19:108-114. ATLA (Alternatives to Laboratory Animals) is published by the Fund for Replacement of Animals in Medical Experiments, Eastgate House, 34 Stoney Street, Nottingham NG1 1NB, England.

Serrano, L. J. 1971. Carbon dioxide and ammonia in mouse cages: Effect of cage covers, population and activity. Lab. Anim. Sci. 21(1):75-85.

Small, J. D. 1987. Drugs used in hamsters with a review of antibiotic-associated colitis. Pp. 179-199 in Laboratory Hamsters, G. L. Van Hoosier, Jr. and C. W. McPherson, eds. Orlando, Fla.: Academic Press.

Snyder, D. L. 1989. Dietary Restriction and Aging. Progress in Clinical and Biological Research, vol 287. New York: Liss.

Society for Research on Biological Rhythms. 1993. Animals issues statement. J. Biol. Rhythms.

Stotzer, V. H., I. Weisse, F. Knappen, and R. Seitz. 1970. Die Retina-Degeneration der Ratte. Arzneim. Forsch. 20:811-817.

Stuhlman, R. A., and J. E. Wagner. 1971. Ringtail in *Mystromys albicaudatus*: A case report. Lab. Anim. Sci. 21:585-587.

Sundstroem, E. S. 1927. The physiological effects of tropical climate. Physiol. Rev. 7:320-362.

Terman, M., C. E. Remé, and A. Wirz-Justice. 1991. The visual input stage of the mammalian circadian pacemaking system: II. The effect of light and drugs on retinal function. J. Biol. Rhythms 6(1):31-48.

Thiessen, D. D. 1964. Population density, mouse genotype and endocrine function in behavior. J. Comp. Physiol. Psychol. 57:412-416.

Thorington, L. 1985. Spectral, irradiance, and temporal aspects of natural and artificial light. Ann. N.Y. Acad. Sci. 453:28-54.

Tobin, R. S. 1987. Testing and evaluating point-of-use treatment devices in Canada. J. Am. Water Works Assoc. Oct., 42-45.

Tobin, R. S., D. K. Smith, and J. A. Lindsay. 1981. Effects of activated carbon and bacteriostatic filters on microbiological quality of drinking water. Appl. Environ. Microbiol. 41:646-651.

Vesell, E. S. 1967. Induction of drug-metabolizing enzymes in liver microsomes of mice and rats by softwood bedding. Science 157:1057-1058.

Vesell, E. S., C. M. Lang, W. J. White, G. T. Passananti, and S. L. Tripp. 1973. Hepatic drug metabolism in rats: impairment in a dirty environment. Science 179:896-897.

Vesell, E. S., C. M. Lang, W. J. White, G. T. Passananti, R. N. Hill, T. L. Clemens, D. K. Liu, and W. D. Johnson. 1976. Environmental and genetic factors affecting the response of laboratory animals to drugs. Fed. Proc. 35:1125-1132.

Wagner, J. E. 1976. Miscellaneous disease conditions of guinea pigs. Pp. 227-234 in The Biology of the Guinea Pig, J. E. Wagner and P. J. Manning, eds. New York: Academic Press.

Wardrip, C. L., J. E. Artwohl, and B. T. Bennett. 1994. A review of the role of temperature versus time in an effective cage sanitation program. Contemp. Top. 33 (5):66-68.

Webb, S. J., R. Bather, and R. W. Hodges. 1963. The effect of relative humidity and inositol on air-borne viruses. Can. J. Microbiol. 9:87-92.

Wegan, R. W. 1982 Alternative disinfection methods—a comparison of UV and ozone. Industrial Water Engineering, March/April, 12-25.

Weihe, W. H. 1965. Temperature and humidity climatograms for rats and mice. Lab. Anim. Care 15(1):18-28.

Weihe, W. H. 1976a. The effects on animals of changes in ambient temperature and humidity. Pp. 41-50 in Control of the Animal House Environment, T. McSheehy, ed. Laboratory Animal Handbooks 7. London: Laboratory Animals Ltd.

Weihe, W. H. 1976b. Influence of light on animals. Pp. 63-76 in Control of the Animal House Environment, T. McSheehy, ed. Laboratory Animal Handbooks 7. London: Laboratory Animals Ltd.

Weindruch, R., and R. L. Walford. 1988. The Retardation of Aging and Disease by Dietary Restriction. Springfield, Ill.: Charles C Thomas.

Weir, B. J. 1967. The care and management of laboratory hystricomorph rodents. Lab. Anim. (London) 1:95-104.

Weir, B. J. 1976. Laboratory hystricomorph rodents other than the guinea-pig and chinchilla. Pp. 284-292 in The UFAW Handbook on the Care and Management of Laboratory Animals, 5th ed, Universities Federation for Animal Welfare, eds. Edinburgh: Churchill Livingstone.

Weisse, I., H. Stötzer, and R. Seitz. 1974. Age- and light-dependent changes in the rat eye. Virchows Arch. A 362:145-156.

West, W. T., and K. E. Mason. 1958. Histopathology of muscular dystrophy in the vitamin E deficient hamster. Am. J. Anat. 102:323.

White, W. J. 1982. Energy savings in the animal facility: Opportunities and limitations. Lab Anim. 2(2):28-35.

White, W. J. 1990. The effects of cage space and environmental factors. Pp. 29-44 in Guidelines for the Well-being of Rodents in Research, H. N. Guttman, ed. Proceedings from a conference organized by the Scientists Center for Animal Welfare and held December 8, 1989, in Research Triangle Park, North Carolina. Bethesda, Md.: Scientists Center for Animal Welfare.

White, W. J., H. C. Hughes, S. B. Singh, and C. M. Lang. 1983. Evaluation of a cubical containment system in preventing gaseous and particulate airborne cross-contamination. Lab. Anim. Sci. 33:571-576.

White, W. J., M. W. Balk, and C. M. Lang. 1989. Use of cage space by guinea pigs. Lab. Anim. (London) 23:208-214.

Williams, F. P., R. J. Christie, D. J. Johnson, and R. A. Whitney, Jr. 1968. A new autoclave system for sterilizing vitamin-fortified commercial rodent diets with lower nutrient loss. Lab. Anim. Care 18:195-199.

Williams, T. P. 1989. Ambient lighting and integrity of the retina. Pp. 75-78 in Science and Animals: Addressing Contemporary Issues, H. N. Guttman, J. A. Mench, and R. C. Simmonds, eds. Bethesda, Md.: Scientists Center for Animal Welfare. Available from SCAW, Golden Triangle Building One, 7833 Walker Drive, Suite 340, Greenbelt, MD 20770.

Williams, T. P., and B. N. Baker, eds. 1980. The Effects of Constant Light on Visual Processes. New York: Plenum Press.

Woods, J. E. 1975. Influence of room air distribution on animal cage enviroments. ASHRAE Trans. 81:559-570.

Woods, J. E. 1978. Interactions between primary (cage) and secondary (room) enclosures. Pp. 65-83 in Laboratory Animal Housing. Proceedings of a symposium organized by the ILAR Committee on Laboratory Animal Housing and held September 22-23, 1976, in Hunt Valley, Maryland. Washington, D.C.: National Academy of Sciences.

Woods, J. E., R. G. Nevins, and E. L. Besch. 1975. Analysis of thermal and ventilation requirements for laboratory animal cage environments. ASHRAE Trans. 81:45-66.

Wurtman, R. J., M. J. Baum, and J. T. Potts, Jr., eds. 1985. The medical and biological effects of light. Ann. N.Y. Acad. Sci. 453:1-408.

Yang, R. S., W. F. Mueller, H. K. Grace, L. Golberg, and F. Coulston. 1976. Hexachlorobenzene contamination in laboratory monkey chow. J. Agric. Food Chem. 24:563-565.

Yu, B. P. 1990. Food restriction: Past and present status. Rev. Biol. Res. Aging 4:349-371.

Yu, B. P., E. J. Masoro, and C. A. McMahan. 1985. Nutritional influences on aging of Fischer 344 rats: I. Physical, metabolic, and longevity characteristics. J. Gerontol. 40:657-670.

Zigman, S., and T. Vaughan. 1974. Near-ultraviolet light effects on the lenses and retinas of mice. Invest. Ophthalmol. Vis. Sci. 13:462-465.

Zigman, S., J. Schultz, and T. Yulo. 1973. Possible roles of near UV light in the cataractous process. Exp. Eye Res. 15:201-208.

Zigman, S., M. Datiles, and E. Torczynski. 1979. Sunlight and human cataracts. Invest. Ophthalmol. Vis. Sci. 18:462-467.

Zimmerman, D. R., and B. S. Wostmann. 1963. Vitamin stability in diets sterilized for germfree animals. J. Nutr. 79:318-322.

Zondek, B., and I. Tamari. 1964. Effect of audiogenic stimulation on genital function and reproduction. III. Infertility induced by auditory stimuli prior to mating. Acta Endocrinol. 45(Suppl. 90):227-234.

6

Veterinary Care

Veterinary care in laboratory animal facilities includes monitoring of animal care and welfare, as well as the prevention, diagnosis, treatment, and control of diseases. It entails providing guidance to investigators on handling animals and preventing or reducing pain and distress. To perform those and related functions, attending veterinarians must be trained or have experience in the care and management of the species under their care. The responsibilities of an attending veterinarian are specified by the Animal Welfare Regulations (AWRs; 9 CFR 2.33 for research facilities and 9 CFR 2.40 for dealers and exhibitors), the *Public Health Service Policy on Humane Care and Use of Laboratory Animals*, or *PHS Policy* (PHS, 1996), and the *Guide for the Care and Use of Laboratory Animals*, known as the *Guide* (NRC, 1996 et seq.).

PREVENTIVE MEDICINE

Procurement

Rodents (excluding mice of the genus *Mus* and rats of the genus *Rattus*) that are acquired from outside a research facility's breeding program must be obtained from dealers licensed by the U.S. Department of Agriculture (USDA) or sources that are exempted from licensing (9 CFR 2.1). Although laboratory mice and rats are excluded from direct USDA oversight, it is recommended that they be acquired from dealers whose facilities and programs conform to the *Guide* (NRC, 1996 et seq.). Documentation of animal health status, site visits by users, history of client satisfaction, USDA licensing for production of other rodent species in the same facilities, and accreditation by the American Association for Accreditation of Laboratory Animal Care can be used to assess dealers.

Sources

Rapid advances in animal-production technology and disease-control methods during the past 20 years have made it easier to obtain laboratory rodents of known health status and genetic definition. Commercial animal producers often maintain colonies of hysterectomy-derived mice, rats, and guinea pigs in barrier facilities designed and operated to prevent the introduction of microbial agents. Those producers regularly monitor their colonies for evidence of infection and infestation and publish the test results in health reports, which they make available to their clients. There is an increasing trend toward maintaining other rodents (e.g., hamsters and gerbils) under similar conditions, although usually not produced from hysterectomy-derived stock. It is recommended that animals be acquired from such sources whenever it is possible and appropriate for the study. When animals that are not barrier-reared are acquired, precautions should be taken to isolate them until health evaluations are conducted and decisions are made regarding their care and use.

Transportation

The protection of the health status of specific-pathogen-free (SPF) rodents during transportation to the user has improved greatly in recent years. USDA supervision of animal carriers has resulted in important changes, including the requirements that rodents covered by the AWRs not be warehoused for long periods before and after shipment, that adequate space be provided in shipping enclosures, and that acceptable temperatures and ventilation be maintained during all phases of transportation (9 CFR 3.35-3.41). The International Airline Transport Association (IATA) has developed guidelines for shipping all animal species, including recommendations for shipping rodents (IATA, 1995 et seq.). Another major improvement has been in the commercial development of disposable shipping containers with filter-protected ventilation openings. In addition, sterile food and moisture sources have become available for use in such containers.

Despite the many changes for the better, problems remain. For example, the potential still exists for contamination of container surfaces during shipment. It is recommended that the surfaces of shipping containers be decontaminated before the containers are moved into clean areas of animal facilities. Several types of disinfectants—including quaternary ammonium solutions, iodinated alcohols, sodium hypochlorite solutions, and chlorine dioxide-containing solutions—can be applied with a small hand sprayer. Chlorine-containing solutions are considered to be very effective against stable agents, such as parvoviruses and spore-forming bacteria (Ganaway, 1980; Orcutt and Bhatt, 1986).

The handling of imported rodents on arrival in U.S. airports can also constitute a problem. Laboratory rodents and rodent tissues that are not inoculated with infectious agents do not require a USDA permit; however, U.S. customs inspectors do not always acknowledge this. Unclear lines of authority often cause unnecessary delays in customs clearance, and such delays can have disastrous effects on the health of the animals. To lessen the probability of delays, as much information as possible should be obtained from the involved authorities (USDA, U.S. Customs, and U.S. Department of the Interior) well in advance of ordering rodents from any foreign source. A permit must also be obtained from the Division of Quarantine, Centers for Disease Control and Prevention, before rodents that can carry zoonotic agents are imported (42 CFR 1, 71.54). Sources of information are listed in the appendix. All necessary documentation should also be obtained before one attempts to export rodents. Specific instructions are usually obtained from the embassy of the country of destination and from the person or institution receiving the animals.

Quarantine and Stabilization

Ideally, rodents being introduced into an animal facility are isolated until their health status can be determined. The period of quarantine also provides time for physiologic and behavioral stabilization after shipment. The users, in cooperation with the veterinarian, should make decisions about the method and duration of quarantine for different kinds of facilities, studies, and types of animals. Unless it is inconsistent with the goals of the study, animals should be allowed to stabilize before the experiment begins.

One of the most common methods of quarantine is to place each group of incoming animals in the same room in which they will eventually be studied. No animals other than those being quarantined should be housed in the quarantine area. For this system to work, each room requires a separate air supply and effective sanitization between studies. Daily animal-care and support activities for quarantine rooms should be conducted after all necessary tasks in the nonquarantine rooms have been performed.

Another approach is to have a single quarantine room for all incoming shipments of animals. This approach has regained favor since the development of isolation-type caging systems, which permit true isolation of many small groups of animals in a single room. Filter-top cages, for example, can be used as miniature rooms within a room. This system works well if animals are moved from dirty to clean cages, one cage at a time in a laminar-flow hood; soiled cages are then closed and autoclaved before they are emptied outside the hood; and appropriate protocols for handling the cages and animals are followed strictly. An advantage of this system is that investigators trained to use it can enter a room and complete short-term studies while the animals are in quarantine. Other variations of quarantine systems have been described elsewhere (NRC, 1991a).

The extent of testing (e.g., serology and parasitology) that is needed during quarantine depends on professional judgment; however, any rodent that dies or becomes ill during quarantine should be subjected to careful diagnostic evaluation. SPF rodents purchased from an established commercial supplier and received in clean, disposable transport cages with filter-protected ventilation openings might not require testing. If the animals are to be used in short-term studies where other short-term studies are performed and relatively few animals are at risk, clinical observations and reliance on the supplier's health program might be adequate. Periodic confirmation of an animal supplier's health report by an independent laboratory provides added safety. If the animals are to be used in a facility where long-term studies might be jeopardized or large numbers of animals are at risk, testing for selected agents of concern is advisable. Maximal protection against the entry of pathogens into a facility is provided by introducing only animals that are delivered by hysterectomy and reared in protective isolation until they are old enough to be tested for the presence of undesirable agents (including agents that can inhabit the female reproductive tract), such as *Mycoplasma pulmonis*, *Corynebacterium kutscheri*, and *Pasteurella pneumotropica*. This course of action is usually followed only in long-standing, ordinarily "closed" breeding colonies.

Animals of undocumented microbiologic status received from any outside source should be serologically tested for a comprehensive list of infectious agents. Animals from such sources might harbor clinically inapparent infectious diseases of major concern. For example, mousepox can be difficult to detect clinically in resistant strains of mice or in mice from colonies with long-standing infections. When introduced into a disease-free colony, mousepox usually becomes evident as an epizootic that can substantially interfere with research (New, 1981). Laboratory rodents and some wild rodents can be subclinically infected with zoonotic agents—e.g., hantaviruses, lymphocytic choriomeningitis (LCM) virus, Lassa fever virus, Machupo virus, and Junin virus—that pose a serious or even deadly health threat to personnel (CDC, 1993; LeDuc et al., 1986; Oldstone, 1987; Skinner and Knight, 1979; Smith et al., 1984). The time of quarantine should be long enough for reasonable expectation that incubating infections will become evident, either clinically or by appropriate testing procedures. As many as

30 percent of the animals should be tested if the microbiologic status of the source colony is completely unknown. In this situation, it is preferable to obtain extra animals for testing so that not only serology, but bacterial cultures, examinations for parasites, and histopathologic evaluations can be performed if needed.

Some pathogens pose special problems for quarantine programs. For example, the chronic form of LCM viral infection in mice, which is contracted in utero or immediately after birth, might not be detectable with antibody tests commonly used in commercial testing laboratories. Mice infected at that time develop persistently high titers of virus that is complexed with humoral antibody, rendering the antibody undetectable by complement-fixation or neutralization tests (Bishop, 1990; Oldstone and Dixon, 1967, 1969). The more-sensitive immunofluorescence assay (IFA) and enzyme-linked immunosorbent assay (ELISA) give weak reactions and cannot be depended on to detect circulating antibody in persistently infected mice (Parker, 1986; Shek, 1994). That is an important problem because the primary route of transmission in the mouse is vertical, and the infected offspring become lifelong, relatively asymptomatic shedders of virus (Rawls et al., 1981). An alternative method for detecting LCM virus in asymptomatic virus shedders is to use virus-free sentinels over the age of weaning (Smith et al., 1984). Once beyond neonatal age, exposed mice develop a short-lived infection and have readily detectable antibodies to LCM virus (Rawls, 1981). Intracranial inoculation of blood or tissue homogenates into the sentinels is a faster screening method. If virus is present, neurologic disease and death will ensue in 6-9 days (Parker, 1986). Additional laboratory procedures would have to be performed to confirm the presence of LCM virus in the dead mice. In testing laboratories that maintain cell lines, such as Vero or BHK-21, the quickest method is to inoculate cell-line cultures with blood from the suspect mice and use the IFA 4-5 days later to test for LCM-virus antigen in the cells. The mouse antibody-production (MAP) test can also be used to detect LCM virus. Antibody to LCM virus in rodents other than persistently infected mice is readily detected with the ELISA or IFA procedures.

Viable rodent tissues—including blood, ascitic fluid, tissue cultures, transplantable tumors, and hybridomas—can harbor undesirable agents, and tissues of undocumented microbiologic status should not be introduced into rodent colonies until they are shown to be free of undesirable agents by diagnostic testing (e.g., MAP testing).

Separation by Species, Source, and Health Status

Pressures to maintain different rodent species in separate rooms have lessened with advances in knowledge of rodent infections. For example, the

AWRs do not require species separation, and the *Guide* (NRC, 1996 et seq.) allows considerable latitude on this issue. It has become recognized that more infectious agents are transmissible among animals of the same species than among those of different species. A more important concern is the microbiologic status of rodents from different sources (or from different locations at the same source), regardless of species. Common sense dictates that if it is necessary to place rodents from different sources in the same room because of space constraints or for other practical reasons, it should be done only with animals of comparable microbiologic status. Such decisions should be made with input from people knowledgeable in rodent-disease pathogenesis and with adequate health-status information about the source colonies.

Interspecies anxiety does not appear to be a problem if different rodent species or rodents and rabbits are housed in the same room, although systematic studies are needed to support the validity of this premise. However, it is unacceptable to house rodents with species that are their natural predators, that produce intimidating noises and odors, or that can harbor infectious agents of known or unknown consequences in rodents (e.g., cats, dogs, and monkeys).

SURVEILLANCE, DIAGNOSIS, TREATMENT, AND CONTROL OF DISEASE

Daily Observations of Animals

One important way to track the health status of rodent colonies is to observe the appearance and behavior of the animals daily. A wide range of abnormal signs can be detected in this manner, including weight loss, ruffled hair coat, dry skin, lacerations, abnormal gait or posture, head tilt, lethargy, swellings, diarrhea, seizures, discharge from orifices, and dyspnea. Underlying causes for those signs include such things as malfunctioning watering systems, fighting, infectious diseases, and experimentally induced changes. Observations are usually made by animal-care staff and technicians, who should be trained to look for spontaneous and experimentally induced abnormalities and report them to the supervisory staff, the attending veterinarian, and study directors. Veterinary oversight of this process and training given by the attending veterinarian are important. Veterinary programs for overseeing the health of laboratory rodents should have readily available, up-to-date references on the biology and diseases of rodents.

Control of Infectious Diseases

First and foremost, control of infectious diseases in rodent colonies means preventing their introduction. That is accomplished by using good

TABLE 6.1 Typical "Core" Agents Monitored in Research Facilities[a]

Agent	Mice	Rats	Guinea Pigs	Hamsters
Kilham rat virus		+		
Minute virus of mice	+			
Mouse hepatitis virus	+			
Mycoplasma pulmonis		+		
Pneumonia virus of mice	+	+	+	+
Rotavirus	+			
Sendai virus	+	+	+[b]	+[b]
Sialodacryoadenitis virus (rat coronavirus)		+		
Simian virus 5			+[b]	+[b]
Theiler's murine encephalomyelitis virus	+			

[a]"Core" agents for each species are indicated by plus signs.

[b]Infection with related parainfluenza viruses can cause false-positive results of tests for Sendai virus and simian virus 5 (Parker et al., 1987).

management practices, such as purchasing pathogen-free animals; using well-planned quarantine systems for incoming animals and animal-derived specimens; training animal-care staff to make accurate clinical observations; using protective clothing; vermin-proofing the facility; using filter-protected cages, filtered-air ventilation systems, or both; and controlling the movement of personnel and visitors within the facility. In addition, animal-care staff should be encouraged not to maintain pet rodents, because of the possibility of transferring infectious agents into the animal quarters.

Even with good management, infections occasionally gain entrance into colonies. Routine monitoring systems should be in place to detect them as quickly as possible, thereby permitting the start of specific measures to eliminate them or prevent their spread. The key elements of an effective monitoring program are daily observation of the animals to detect clinical diseases and regular microbiologic monitoring to detect subclinical infections. Daily observations are extremely important because they quickly reveal signs of spontaneous disease. To achieve full effectiveness, monitoring activities require diagnostic capability to investigate disease outbreaks.

Microbiologic monitoring can include many kinds of tests, depending on the needs of the facility. Animal suppliers often test for all infectious agents of rodents for which there are commercially available tests so that fully characterized animals can be offered for research use. In research facilities, the staff might choose to test initially or annually for all known pathogenic agents and test more frequently for a smaller number of "core" agents of special concern. Table 6.1 lists typical "core" agents. The research requirements or special interests of the staff will dictate what other agents should be added to the list.

Several newly recognized viruses that are not listed as core agents deserve mention because of their apparent high prevalence. These are the so-called orphan parvoviruses of mice and rats that appear to be widespread in laboratory colonies but are of unknown character and pathogenicity. Although field strains of the viruses are yet to be isolated, the mouse orphan parvovirus (MOPV) has been demonstrated in tissues by in situ hybridization (Smith et al., 1993), and a closely related laboratory strain has been isolated (McKisic et al., 1993). In routine testing, the viruses of both mice and rats have been detected indirectly by IFA demonstration of antibody against nonstructural proteins of the rodent parvovirus group followed by negative results with hemagglutination inhibition (HAI) tests that are specific for recognized parvoviruses (i.e., MVM, KRV, and Toolan H-1 virus). An HAI test specific for MOPV has been developed by using the laboratory strain (Fitch isolate) but is not yet in general use.

It is debatable whether Sendai virus and simian virus 5 (SV5) should continue to be listed as core agents for guinea pigs and hamsters. Although serologic positivity is often found, it is believed by some to be caused by infection with antigenically related parainfluenza viruses, possibly from human sources. Isolation of Sendai virus from guinea pigs has been attempted rarely and described only anecdotally (Parker, reported by Van Hoosier and Robinette, 1976). Failure of transmission of Sendai virus from serologically positive guinea pigs to mice also has been found (W. White, Charles River Laboratories, Wilmington, Massachusetts, unpublished). Isolation of Sendai virus from hamsters has been reported rarely (Parker et al., 1987). Serologic positivity for Sendai and SV5 viruses might be caused by cross reactions with human parainfluenza viruses, but isolation of the human agents from these animals has not been documented.

Monitoring can be performed for many combinations of agents and with various frequencies. Emphasis is often on serologic testing because many of the agents of concern cause subclinical infections and are detectable quickly and inexpensively with this method. Table 6.2 lists infectious agents of commonly used laboratory rodents for which serologic (antibody) tests are available.

Bacteriologic testing usually entails culturing for primary and opportunistic pathogens from the upper respiratory tract and intestines. Table 6.3 lists the primary pathogens culturable from these sites.

Monitoring for ectoparasites is done usually by examining the skin and pelage over the head and back with a dissection microscope. For parasites that invade the skin, skin scrapings in immersion oil or 5 percent potassium hydroxide are examined microscopically. Monitoring for endoparasites is performed by using fecal flotation and sedimentation procedures to search for eggs and oocysts, using the Cellophane-tape method to look for *Syphacia* eggs, examining the cecocolic contents for helminths, and examining the blad-

TABLE 6.2 Infectious Agents of Rodents for Which Serologic Tests Are Available

	Serologic Test Available[a]			
Agent	Mice	Rats	Guinea Pigs	Hamsters
Clostridium piliforme (formerly called *Bacillus piliformis*)	+	+		
Cilia-associated respiratory (CAR) bacillus	+	+		
Ectromelia virus	+			
Encephalitozoon cuniculi	+	+	+	
Hantavirus	+	+		
K virus	+			
Kilham rat virus		+		
Lymphocytic choriomeningitis virus	+		+	+
Minute virus of mice	+			
Mouse adenovirus (MAd-FL, MAd-K87)	+	+		
Mouse cytomegalovirus	+			
Mouse hepatitis virus	+			
Mouse "orphan" parvovirus	+			
Mouse rotavirus	+			
Mouse thymic virus	+			
Mycoplasma arthritidis	+	+		
Mycoplasma pulmonis	+	+		
Pneumonia virus of mice	+	+	+	+
Polyoma virus	+			
Rat coronavirus and sialodacryoadenitis virus		+		
Rat cytomegalovirus		+		
Rat "orphan" parvovirus		+		
Reovirus 3	+	+	+	+
Sendai virus	+	+	+	+
Simian virus 5			+	+
Theiler's murine encephalomyelitis virus	+	+		
Toolen's H-1 virus		+		

[a]Agents for which serologic tests are available are indicated by plus signs.

der mucosa for *Trichosomoides crassicauda* (in rats) and fecal wet smears for protozoa. Descriptions of ectoparasites and endoparasites and their effects on rodents have been published (Farrar et al., 1986; Flynn, 1973; Hsu, 1979, 1982; Ronald and Wagner, 1976; Vetterling, 1976; Wagner, 1987; Wagner et al., 1986; Weisbroth, 1982; Wescott, 1976, 1982). Pathologic monitoring can be used to detect diseases that produce characteristic lesions that are observable at necropsy or detectable by histopathologic evaluation. Infectious diseases for which this approach is useful include Tyzzer's disease (*Clostridium piliforme* [formerly called *Bacillis piliformis*] infection), pneumocystosis (*Pneumocystis carinii* infection) in some immunodeficient animals, and CAR

TABLE 6.3 Important Rodent Bacterial Pathogens Culturable from Upper Respiratory Tract and Intestines[a]

Agent	Mice	Rats	Guinea Pigs	Hamsters	Gerbils
Bordetella bronchiseptica			+		
Campylobacter jejuni				+	
Citrobacter freundii (biotype 4280)	+				
Corynebacterium kutscheri	+	+		+	
Helicobacter spp.	+				
Mycoplasma pulmonis	+	+			
Salmonella spp.	+	+	+	+	+
Streptobacillus moniliformis		+			
Streptococcus equis (*zooepidemicus*)			+		
Yersinia pseudotuberculosis			+		

[a]Culturable pathogens are indicated by plus signs. Many commonly occurring bacteria can be present as pathogenic strains (e.g., *Escherichia coli* and *Streptococcus pneumoniae*) or as opportunistic pathogens (e.g., *Klebsiella* spp., *Pasteurella pneumotropica*, and *Pseudomonas aeruginosa*) in stressed or immunocompromised animals, or as agents of importance when transmitted from a carrier to a susceptible animal host (e.g., *Bordetella bronchiseptica*).

bacillus infections. Special stains are required to demonstrate those causative agents (e.g., methenamine silver for *P. carinii* and Warthin Starry silver for *C. piliforme* and CAR bacillus). Pathologic monitoring can also be used to detect noninfectious conditions, such as nutritional deficiencies, heritable metabolic diseases, and neoplasms. The necropsy is usually the first step in the diagnostic workup of clinical diseases, often providing the impetus for using other measures, such as virus isolation, bacterial cultures, or histopathology. Complete descriptions of these procedures and the manifestation of infections in rodents are beyond the scope of this report, but such information is available in a number of books, manuals, and review articles (ACLAD, 1991; Baker et al., 1979; Bhatt et al., 1986; Flynn, 1973; Foster et al., 1982; Hamm, 1986; NRC, 1991a; Van Hoosier and McPherson, 1987; Waggie et al., 1994; Wagner and Manning, 1976).

Sample Size for Monitoring

All animals should be monitored for clinical disease by daily observations. This type of monitoring, combined with a diagnostic workup of animals with unexplained abnormalities, is particularly important for early detection of clinical disease outbreaks. It is complementary to microbiologic monitoring in that diseases that spread slowly and smolder for a considerable time in a few cages in a room (Bhatt and Jacoby, 1987; Wallace et al., 1981) might be missed in the statistical sampling used in microbiologic monitoring. Daily observations should quickly reveal these kinds of diseases.

Microbiologic monitoring for evidence of subclinical infections is accomplished by testing regularly a randomly selected sample of the population of animals at risk. How to determine the appropriate sample size is a much debated subject. A formula has been used to predict the number of randomly selected animals in a population of 100 or more that must be tested to detect a single case of disease within 95 percent confidence limits, assuming a known prevalence rate (NRC, 1976):

$$\text{No. to be sampled} = \frac{\log 0.05}{\log N}.$$

In that formula, N is the percentage of animals expected to be normal. The percentage is derived by subtracting the expected prevalence rate of the disease from 100 percent. The formula is useful for helping to understand the considerations involved in sampling to detect a single disease. In practice, however, its use is limited by several factors. One factor is that sampling of a rodent population is usually aimed at detecting more than one disease, each with a different expected prevalence. Another problem is that infectious-disease prevalences are affected by population density, caging methods, ventilation systems, and a host of other variables that affect the rate of spread of infections; a disease prevalence expected to be 30 percent in open cages might be only 1 percent in filter-top cages. Still another consideration is that much of the monitoring is done by testing for antibody. If an infection with an expected prevalence of 30 percent has been in a colony for several months, the number of surviving animals with antibody can approach 100 percent. Because of those variables, the formula serves only as a rough estimate. If it is used, one prevalence is selected for all diseases and conditions, even though screening is usually for multiple organisms. For example, a prevalence of 30 percent might be assumed for more contagious infections, and a sample size of 8-10 would be used. This sample size would, of course, be unlikely to detect infections that are less contagious (NRC, 1991a).

Similar calculations can be made for populations of fewer than 100 with other formulas. More complex calculations can be used once the monitoring program is in place and sufficient data have been accrued on the incidence of positive findings and frequency of disease outbreaks. Those calculations can be used to adjust the sample size and frequency of sampling to achieve the desired confidence levels for disease detection (Selwyn and Shek, 1994).

In summary, there is no easy way to determine sample sizes and frequencies for monitoring. Although a mathematical approach can be taken, the inability to conform to the assumptions on which the formulas are based or the lack of precise knowledge of prevalence rates or disease outbreaks makes such an approach difficult to apply. For that reason, it is still common to choose sample size and frequency of monitoring in an arbitrary manner, which is often influenced by economic constraints.

An alternative method of monitoring uses *known* pathogen-free sentinel animals to detect infections. Typically, they are randomly dispersed in multiple locations in the facility, and various means are used to promote contagion of any infections that might be present from the animals being monitored by the sentinels. The most effective method is to place the sentinels in the cages with the study animals and move them to cages of different study animals every 1-2 weeks. If such a procedure is not practical, the sentinels should at least be caged on the same rack with the study animals, preferably on a lower shelf, and soiled bedding from the cages of the study animals should be transferred regularly to the cages of the sentinel animals (Thigpen et al., 1989). Because natural transmission of some pathogens might not occur quickly, the time allowed for seroconversion or production of disease should be about 6-8 weeks. Those pathogens include *Mycoplasma pulmonis* (Cassell et al., 1986; Ganaway et al., 1973), ectromelia virus (Wallace et al., 1981), and cilia-associated respiratory (CAR) bacillus (Matsushita et al., 1989); a preferable alternative is to test the animals being introduced into the colony rather than the sentinels.

Treatment and Control

Health-monitoring data should be reviewed regularly, and a plan of action should be in place for dealing with positive test results. Such plans usually include the names and telephone numbers of research and veterinary staff to be notified, a system for confirming the test results, and appropriate measures for controlling or eliminating infection. Decisions about ways to prevent spread to contiguous areas should be made quickly. They usually involve placing the room under strict quarantine and developing strategies for controlling access and for handling potentially contaminated items, such as cages and bedding, that will be removed from the room periodically. Investigations are usually initiated immediately to identify the sources of causative agents. Approaches to control depend on the characteristics of the agents, the value of the infected animals, and the type and design of the facility.

Bacterial diseases of rodents can be treated with antibiotics. However, when large numbers of animals are involved, this is often considered practical only for temporary control. Failure to eliminate the agent from every animal, as well as from contaminated surfaces, might result in re-emergence of the disease when antibiotics are discontinued. In some instances, antibiotics can adversely affect rodents, especially guinea pigs and hamsters, by causing an imbalance of the intestinal microflora and overgrowth of deleterious bacteria (Fekety et al., 1979; Small, 1968; Wagner, 1976). Other problems include the lack of information on proper dosages, the difficulty of accurately administering antibiotics in food and water, and confounding influences of drug residues and interactions on research results.

Parasitic diseases can also be treated; however, even with highly effective antiparasitic drugs, it is very difficult to eliminate from large colonies such parasites as pinworms and mites. It might be possible in small colonies if the treatment schedule is adjusted to overlap the time of the parasite life cycle and if sanitation procedures are stringently performed simultaneously (e.g., frequent washing of floors, walls, and cages) (Findon and Miller, 1987; Flynn et al., 1989; Silverman et al., 1983; Taylor, 1992; West et al., 1992).

Viral, bacterial, and parasitic infections are usually eliminated by euthanatizing and repopulating the colony with disease-free animals after the room, cages, and other equipment have been decontaminated or, in the case of particular viruses, by allowing the infection to run its course in a closed population to produce noninfected, immune survivors. The latter procedure has been used successfully with such viruses as Sendai virus and mouse hepatitis virus, which are highly contagious, usually remain in the animals for a short time, and are relatively unstable in the animal-room environment (Barthold, 1986; Fujiwara and Wagner, 1986). For it to be successful, ample opportunity for contagion is required, and new animals, even newborns, must not be introduced for a period long enough for all animals to become infected, recover, and stop shedding the virus. Contagion can be promoted by transferring infected bedding to numerous cages, placing cage racks near each other, and removing filter tops. Sentinels can be introduced and tested 6-8 weeks later to determine the success of the procedure. No sentinels should be introduced into the room, and no naive animals of any type should be allowed to be introduced or maintained in the room until 6-8 weeks after breeding has been stopped.

Necropsies

When an animal is unexpectedly found dead or moribund, it is good practice to determine the cause by necropsy. Necropsy, coupled with daily observations by the animal technicians, usually provides the first indication of important clinical infectious and noninfectious diseases. Lesions will often be characteristic enough to permit presumptive diagnoses or point to appropriate additional diagnostic procedures. Routine histopathologic tests are performed in some facilities.

EMERGENCY, WEEKEND, AND HOLIDAY CARE

The need for adequate animal care does not diminish during holidays and weekends. As stated in the *Guide*, laboratory animals should be cared for daily (NRC, 1996 et seq.) Security personnel should be able to contact responsible people in the event of emergencies. Therefore, a list of names and phone numbers should be posted prominently in the facility and maintained in the security office. Provisions for emergency veterinary care should be made as well (9 CFR 2.33b2; NRC, 1996 et seq.).

MINIMIZATION OF PAIN AND DISTRESS

Many internal and external environmental factors can induce physiologic or behavioral changes in laboratory animals. These factors are called stressors, and their effect is called stress (NRC, 1992). The intensity of the stress experienced by an animal is influenced by other factors, including age, sex, genetics, previous exposure, health status, nutrition, and medication (Blass and Fitzgerald, 1988; NRC, 1992). If an animal is unable to adapt to stressors, it will develop abnormal physiologic or behavioral responses; when this occurs, the animal is in distress (NRC, 1992). Sometimes, the effect induced by the stressor is pain. Pain can be described as a physical discomfort perceived by an organism as the result of injury, surgery, or disease. Once pain is perceived by an animal, it can itself become a secondary stressor and elicit other responses, such as fear, anxiety, and avoidance.

To prevent or alleviate pain and distress in laboratory rodents, the research team should anticipate procedures or situations that will elicit these conditions. According to the *U.S. Government Principles for the Utilization and Care of Vertebrate Animals Used in Testing, Research, and Training*, "unless the contrary is established, investigators should consider that procedures that cause pain or distress in human beings may cause pain and distress in other animals" (published in NRC, 1996, p. 82). Classifications of the magnitude of pain or distress estimated to be associated with different types of experimental procedures are available in the literature (NRC, 1992; OTA, 1986). It is the responsibility of the institutional animal care and use committee (IACUC) to evaluate each animal procedure for the potential to cause pain or distress and to ensure that anesthetics, analgesics, and tranquilizers are used, when appropriate, to prevent or alleviate pain and distress in the animals. Anesthetics or analgesics should be given before the painful insult, because it is easier to prevent pain, by blocking nociceptive neurons, than to alleviate it. The exposure of nociceptive neurons to painful stimuli produces chemical changes that cause the neurons to be hypersensitive to additional pain stimuli for a long period (Hardie, 1991; Kehlet, 1989). In addition, a cascade of physiologic changes occur that can have substantial effect on the recovery of an animal from surgery or on the information that is obtained in the procedure in which the animal is used. Depending on whether the pain is acute or chronic, responses might include protein catabolism, sodium retention, immunosuppression, decreases in pulmonary and cardiovascular function, and increases in plasma concentrations of catecholamines and corticosteroids (Engquist et al., 1977; Flecknell, 1987; S. A. Green, 1991; Yeager, 1989).

Recognition of Pain and Distress

Every person involved in the procurement, care, and use of laboratory rodents plays a major role in contributing to the total well-being of these animals. It is important to understand and consider species-specific behavior and husbandry needs when standard operating procedures and research protocols are developed to minimize exposure of the animals to situations that have a high probability of inducing pain and distress (Amyx, 1987; Montgomery, 1987).

Clinical signs and abnormal behavior displayed by rodents in response to pain and distress can include decreases in food and water consumption, accumulation of reddish-brown exudate around the eyes and nostrils (chromodacryorrhea), weight loss, decrease in activity, hunched posture, piloerection, poor grooming habits, labored respiration, vocalization, increase or decrease in aggressiveness, and self-mutilation (Flecknell, 1987; Flecknell and Liles, 1992; Harvey and Walberg, 1987; Heavner, 1992; NRC, 1992; Sanford, 1992). The degree to which clinical signs are displayed varies within a species and between species. For behavior to be a useful indication of pain or distress, members of the research team, from animal caretakers to principal investigators, should be knowledgeable about the normal behavior of the animals with which they are working. Regular communication among all members of the research team, including the veterinary staff, is critical to ensuring timely evaluation and treatment of animals in pain or distress.

Alleviation of Pain

The *Guide* recommends the use of appropriate anesthetics, analgesics, and tranquilizers for the prevention and control of pain and distress. However, if for justifiable scientific reasons these agents cannot be administered when a painful procedure is to be conducted, the *Guide* states that the procedure must be approved by the committee [IACUC] and conducted by persons with adequate training and experience in the procedure used (NRC, 1996, p.10).

The drugs routinely used to prevent or control pain in laboratory rodents are generally classified as either opioids or nonsteroidal anti-inflammatory agents. Drugs reported to be effective analgesics in rodents are published elsewhere (Blum, 1988; CCAC, 1980; Clifford, 1984; Flecknell,

1984, 1987; C. J. Green, 1982; Hughes, 1981; Hughes et al., 1975; Jenkins, 1987; Kruckenburg, 1979; Lumb and Jones, 1984; Soma, 1983; Vanderlip and Gilroy, 1981; White and Field, 1987). In some cases, the doses quoted are extrapolations from doses for other species, with little or no scientific evidence to support the recommended use. Because some of these drugs might have systemic side effects that could interfere with a research protocol, it is important to select and use them carefully. Additional factors that should be considered in selecting an analgesic include species, strain, age, sex, health status, nutritional status, period for which pain prevention or control will be required, recommended route of administration, volume of drug required for effect, compatibility with other pharmacologic agents that the animal will be receiving, cost, and availability (C. J. Green, 1982; Kanarek et al., 1991; Pick et al., 1991). Principal investigators should get assistance from the attending veterinarian in selecting the most appropriate agent.

Alleviation of Stress and Distress

The use of tranquilizers can be considered when a laboratory rodent is restrained for long periods or used in a procedure that might cause fear, anxiety, or severe distress. Dosages of tranquilizing agents for rodents have been reported elsewhere (Blum, 1988; CCAC, 1980; Flecknell, 1987; C. J. Green, 1982; Harkness and Wagner, 1989; NRC, 1992; Vanderlip and Gilroy, 1981; White and Field, 1987). It should be noted, however, that tranquilizers have not been well studied in rodents. The drugs might interfere with experimental results, and suggested dosages might not produce the desired effects. Gradual conditioning to restraint before initiation of a study should also be considered as a means of decreasing associated anxiety or distress.

SURVIVAL SURGERY AND POSTSURGICAL CARE

Surgical procedures on rodents must be performed only by appropriately trained personnel or under the direct supervision of trained personnel (9 CFR 2.32; NRC, 1996 et seq., 1991b). It is essential that personnel given the responsibility to perform surgery be knowledgeable about the principles of aseptic technique and the correct methods for handling tissues and using surgical instruments (McCurin and Jones, 1985). It is the responsibility of the IACUC to ensure that people approved to perform surgery on rodents have the required training or experience (9 CFR 2.32).

Standards and guidelines for conducting survival surgery have been established by the *Guide* (NRC, 1996 et seq.) and for rodents other than mice and rats by the AWRs (9 CFR 2.31). Aseptic technique is required whenever a *major* survival surgical procedure is performed. Aseptic technique is used to reduce microbial contamination to the lowest practical level

(Cunliffe-Beamer, 1993) and includes preparation of the animal, preparation of the surgeon, sterilization of instruments and supplies, and the use of operative procedures that reduce the likelihood of infection. A major surgical procedure has been defined as any surgical intervention that penetrates a body cavity or produces permanent impairment of physical or physiologic function (9 CFR 1.1; NRC, 1996 et seq.). Other surgical procedures, classified as *minor*, include catheterization of peripheral vessels and wound suturing. Less stringent conditions are permitted for minor surgical procedures (NRC, 1996, p. 62), but sterile instruments should be used and precautions should be taken to reduce the likelihood of infection. Deviations from those guidelines and standards should not be undertaken unless reviewed and approved by the IACUC.

The susceptibility of rodents to surgical infection has been debated; however, available data suggest that subclinical infections can cause adverse physiologic and behavioral responses (Beamer, 1972-1973; Bradfield et al., 1992; Cunliffe-Beamer, 1990; Waynforth, 1980, 1987), which can affect both surgical success and research results. Characteristics of surgery on rodents that can justify modifications in standard aseptic technique include smaller incision sites, multiple operations at one time, shorter procedures, and complications caused by the use of antibiotics (Brown, 1994; Cunliffe-Beamer, 1993; Small, 1987; Wagner, 1976). Strategies have been published that provide useful suggestions for dealing with some of the unique challenges of rodent surgery (Cunliffe-Beamer, 1983, 1993). The area used for surgery, whether or not it is dedicated for that use, must be easily sanitized, must not be used for any other purpose during the time of surgery, and should be large enough to enable the surgeon to conduct the procedure without breaking aseptic technique.

It might be necessary to perform experimental surgery on animals whose health has been compromised by naturally occurring or experimentally induced disease, but generally only healthy rodents should be used in experimental surgical procedures. Before being used in experimental surgery, rodents should be allowed sufficient time to acclimate to a new environment and overcome the stress of transportation. Results of several studies have shown that mice experience increased corticosterone concentrations and depressed immune function after transport; these functions return to baseline values within 4-8 hours. The length of time might vary with the species and the mode and duration of transportation (Aguila et al., 1988; Dymsza et al., 1963; Landi et al., 1982; Selye, 1946). During the acclimation period, the animals should be examined to ensure that they are not exhibiting clinical signs of disease.

To reduce or prevent stress preoperatively, researchers should be trained to handle and restrain animals and give them injections properly (NRC, 1991b). The animals should be conditioned to being picked up and handled by the people that will be doing the preoperative procedures. Fasting for periods of 12 hours or more is neither recommended nor generally required. However, it is often desirable to remove food at least 4 hours before anesthesia to promote consistent absorption of intraperitoneal anesthetics (White and Field, 1987). Access to water should be allowed up to the time that preoperative procedures are to begin (C. J. Green, 1982).

Anesthetics and Tranquilizers

Administration of tranquilizers, sedatives, or anesthetics might prevent or alleviate stress in the animals, as well as making it easier for surgical personnel to prepare them for surgery. Dosages of tranquilizers and anesthetics that can be used in rodents have been reported elsewhere (Blum, 1988; Flecknell, 1987; C. J. Green, 1982; Harkness and Wagner, 1989; Hughes, 1981; Kruckenburg, 1979; Soma, 1983; Stickrod, 1979; White and Field, 1987). In addition to injectable and inhalational anesthetics, hypothermia has been recommended as a means of anesthesia in neonatal animals (C. J. Green, 1982; NRC, 1992; Phifer and Terry, 1986). Criteria for selecting tranquilizers and anesthetics and their dosages should include species, strain, age, sex, health status, temperament, environmental conditions of the animal holding rooms, drug availability, drug side effects, recommended route of administration, equipment required, length of time that drug effect is desired, and skills and experience of the anesthetist. Doses quoted are often extrapolations from doses for other species with little or no scientific evidence to support them. It is important to select and use these drugs carefully to avoid interference with research protocols.

Preparation for Survival Surgery

Once the animal is tranquilized, sedated, or anesthetized, the operative site should be prepared. The extent of this preparation will depend on the species and maturity of the animal and on the complexity of the surgical procedure to be performed. The preparation might include removing body hair along the surgical site and surrounding areas with clippers, razors, or depilatory agents or by manual plucking. Care should be taken to avoid physical or chemical damage to the skin. Loose hairs should be thoroughly cleared from the surgical site. Various commercially available agents are appropriate for disinfecting the skin, including povidone iodine, alcohol, and chlorohexidine. Because the blink reflex is often lost under general anesthesia, consideration should be given to applying a sterile ophthalmic lubricant before surgery to prevent drying of the corneas (Powers, 1985).

Heat loss can affect the course and success of anesthesia in rodents. Rodents lose body heat rapidly to surfaces such as operating tables, bench tops, and instruments. To preserve body heat, a circulating hot-water blanket, hot-water bottles, or an incandescent lamp placed 12-14 inches from the animal can be used to supply supplemental heat during the surgical procedure and recovery. Positioning the animal on an insulating surface, such as cloth or paper, will also help to decrease heat loss.

The animal should be positioned to provide adequate fixation and exposure of the operative site. Tape, positional ties, or similar mechanical means should be used to ensure that the animal's position will not be changed by pressure exerted by the surgeon. Care should be taken so that the selected method of restraint does not impede circulation or cause injury to the animal.

Depending on the complexity of the surgical procedure, it might be necessary to place a sterile drape over the animal to prevent contamination of the operative site. Various commercially available cloth, paper, and plastic materials are suitable for use as surgical drapes.

In preparation for the procedure, the surgeon should scrub his or her hands and forearms with a povidone iodine scrub, alcohol foam product, or other equally effective disinfectant-detergent. At a minimum, surgical personnel must wear sterile gloves while performing surgery (9 CFR 2.31; NRC, 1996 et seq.). For rodents other than mice of the genus *Mus* and rats of the genus *Rattus*, masks are also required by the AWRs (9 CFR 2.31). Although caps and gowns are not required for rodent surgery, their use can decrease the risk of contaminating the surgical site and sterile supplies.

Sterilization of Instruments

The AWRs (9 CFR 2.31) and the *Guide* (NRC, 1996 et seq.) require that all instruments used in survival surgery be sterilized. As many sets of sterilized instruments as possible should be available when a surgical procedure will be performed on multiple animals during the same operative period. If it is necessary to use the same instruments on several animals, instruments that were sterile at the beginning of the procedure should, at a minimum, be disinfected by chemical or other means (e.g., heated glass beads) before they are used on another animal.

Various methods and materials are available for sterilization of instruments and surgical supplies, including heat, steam under pressure, ethylene oxide gas, gamma irradiation, electron-beam sterilization, and such chemical agents as phenols and glutaraldehyde. The method selected should be periodically monitored (e.g., with spore strips in autoclaves) to ensure that sterilization is achieved. When ethylene oxide gas or a liquid chemical agent is used, care should be taken to ensure that all toxic residues are eliminated before the instruments and supplies are used for surgical procedures.

Instruments and supplies that are to be sterilized with methods other than contact with liquid agents should be wrapped in paper, cloth, plastic, or similar materials in such a way as to prevent contamination after sterilization. The choice of material should be appropriate for the method of sterilization. Each package should bear some indication that it has undergone sterilization. The package should also be marked with the date of sterilization. The shelf-life of sterilized items will depend on the type of material used to wrap them and on how they are stored (Berg and Blass, 1985; Gurevich, 1991; Knecht et al., 1981). Items that are sterilized with liquid agents are generally prepared near the operating room or area and used immediately after they are removed from the liquid and rinsed with sterile water or sterile irrigation solution.

Monitoring During Surgery

Surgical procedures should not be initiated until the animal has reached a surgical plane of anesthesia. In most rodents, loss of toe-pinch and pedal reflexes indicates that the plane of anesthesia is adequate for surgery. Guinea pigs, however, can maintain a pedal reflex under anesthesia; for them, the pinna reflex is more appropriate for assessing the plane of anesthesia (C. J. Green, 1982). The animals should be closely monitored throughout the procedure. An animal's status can be determined by monitoring respiration, eyes, and mucous membranes. Slow, labored respiration, loss of reflected eye color in albino animals, and pale or cyanotic mucous membranes are all indicators of compromised cardiovascular and respiratory functions. If resuscitation is necessary, a modified bulb syringe can be fitted over the animal's muzzle and gently pumped to force air into its lungs. A gentle, rhythmic pressure can be applied over the apical area of the thorax to induce cardiac contractions. Doxapram can be used to stimulate respiration (Flecknell, 1987). The attending veterinarian can instruct investigators about those and other resuscitative techniques most appropriate for the species and procedures used.

Postoperative Care

A rodent recovering from surgery should be observed regularly until it is conscious and has regained its righting reflex. It should be housed singly in a cage on absorbent material that minimizes heat loss until it is conscious. Recovery is facilitated by providing supplemental heat as previously described. Care should be taken to prevent thermal injuries if water bottles, electric heating pads, or heating lamps are used.

If necessary, body fluid lost during the surgical procedure should be replaced with subcutaneously or intraperitoneally administered fluids. A decision to administer fluids should be based on the nature and length of the surgical procedure and an estimation of fluid loss. Sterile saline, lactated Ringer's and 5 percent glucose solutions are often used. Guidelines on fluid-replacement therapy are available (Cunliffe-Beamer and Les, 1987; Lumb and Jones, 1984).

If recovery takes longer than 30 minutes, the animal's position should be rotated to prevent congestion in dependent organs. If there is concern that its toes will become entangled in sutures or that it will harm the incision or damage the bandage or other protective devices, its toenails should be clipped during the postoperative recovery period.

Analgesics should be administered as needed during the postoperative recovery period. Possible side effects and drug interactions should be taken into consideration when specific agents are selected for use (Harkness and Wagner, 1989).

Surgical wounds should be examined daily for dehiscence, drainage, and signs of infection. Appropriate nursing care should be given to prevent drainage from the incision from irritating the surrounding skin. If nonabsorbable sutures or medical staples are used to close the skin, they should be removed when the incision is adequately healed.

EUTHANASIA

Euthanasia is the act of producing a painless death. It entails disrupting the transmission of signals from peripheral pain receptors to the central nervous system (CNS) and rendering the cerebral cortex, thalamus, and subcortical structures of the CNS nonfunctional. The "endpoint" (the point at which euthanasia will be performed) should be specified in any protocol for a terminal study or for a study in which the animals are likely to experience pain and distress that cannot be adequately controlled or prevented with pharmacologic agents, including studies associated with infectious diseases or tumor growth. Each investigator should consult with the attending veterinarian to decide on a humane endpoint that will allow collection of the required data without causing undue pain and distress (Amyx, 1987; Montgomery, 1987).

The technique selected for performing euthanasia on laboratory rodents should be based on a number of factors, including the following:

- species;
- animal age and condition;
- objectives of the study;
- histologic artifacts and biochemical changes induced by the agent or method selected;
- number of animals to be euthanatized;
- available personnel;
- cost and availability of supplies and equipment;
- controlled-substance use; and
- skills of assigned personnel.

To avoid causing stress in the animals that will be euthanatized, the following principles should be adhered to:

- Animals should not be euthanatized in the same room in which other animals are being held. The visual, acoustic, and olfactory stimulants that can be present at euthanasia can cause distress in other animals.
- Animals should be handled gently and humanely during transport from the holding room and during the actual euthanasia process.
- If a euthanasia chamber is used, overcrowding should be avoided.
- Euthanasia should be performed only by people trained in the method selected. It is important that the training received include basic information on how the technique works to produce a quick and painless death and on the advantages of using a specific method in a specific protocol.
- Counseling should be available for those performing euthanasia to help them understand feelings and reactions that might develop as a result of performing this task.
- Death should be verified at the end of the procedure. Possible methods might include exsanguination, decapitation, creation of a pneumothorax by performing a bilateral thoracotomy or incising the diaphragm, and a physical examination to verify the absence of vital signs.

PHS Policy (PHS, 1996) requires that methods of euthanasia be consistent with the recommendations of the American Veterinary Medical Association (AVMA) Panel on Euthanasia (AVMA, 1993 et seq.). AVMA-recommended methods cause death by direct or indirect hypoxia, direct depression of CNS neurons, or physical damage to brain tissues. The approved pharmacologic agents and physical methods include barbiturates, inhalant anesthetics, carbon dioxide, carbon monoxide, nitrogen, argon, and microwave irradiation. Two additional techniques, cervical dislocation and decapitation, can be used if scientifically justified and approved by the IACUC (AVMA, 1993). Of these agents and methods, four are commonly used for rodents: carbon dioxide, sodium pentobarbital, cervical dislocation, and decapitation.

Carbon dioxide is a very safe and inexpensive agent for euthanatizing laboratory rodents. In all but neonates, it causes rapid, painless death by a combination of CNS depression, which is produced by a fall in the pH of the cerebrospinal fluid, and hypoxia. Other methods of euthanasia can be used in newborn animals, which are more resistant to acute respiratory acidosis and hypoxia than older animals. Commercially available cylinders of compressed carbon dioxide or blocks of dry ice can used as the source of carbon dioxide. Compressed gas is preferable because inflow to the chamber can be regulated precisely (AVMA, 1993). If dry ice is used, it should be placed in the bottom of the chamber and separated from the rodent by a barrier to prevent direct contact that could cause chilling or freezing and associated stress.

Sodium pentobarbital is the barbiturate drug most commonly used for euthanatizing animals and can be administered to rodents either intraperitoneally or intravenously. When administered intravenously to rodents at a dose of 150-200 mg/kg of body weight (NRC, 1992), it causes rapid death by CNS depression and hypoxia. Intracardiac and intrapulmonary routes of administration can cause pain and distress because of the required methods of restraint and other procedural difficulties. Therefore, those routes of administration should not be used unless the animal is anesthetized.

Cervical dislocation is an acceptable method for euthanatizing rodents, provided that it is performed by appropriately trained personnel. Death is instantaneous and is caused by physical damage that occurs as the brain and spinal cord are manually separated by anteriorly directed pressure applied to the base of the skull. This technique might be more difficult to perform in hamsters, rats, and guinea pigs than in other rodents because of the strong muscles and loose skin of the neck region. If the method is selected, it should be remembered that it can produce pulmonary artifacts—blood in the alveoli and vascular congestion (Feldman and Gupta, 1976).

For decapitation, only a sharp, clean guillotine or large shears should be used to ensure a clean cut on the first attempt. It is also essential that the cut be made between the atlanto-occipital joint to ensure that all afferent nerves are severed (NRC, 1992). Decapitation is more difficult in hamsters, rats, and guinea pigs than in other rodents because of the strong muscles and loose skin of the neck region. There has been considerable controversy about how rapidly unconsciousness occurs when this method is used and whether animals should be anesthetized before they are decapitated. There is evidence that unconsciousness occurs very rapidly (in less than 2.7 seconds) after decapitation (Allred and Berntson, 1986; Derr, 1991). Recent studies have shown that anesthesia can cause substantial alterations in arachidonic acid metabolism; lymphocyte assays; and plasma concentrations of glucose, triglycerides, and insulin (Bhathena, 1992; Butler et al., 1990; Howard et al., 1990). It can be concluded that in some cases anesthesia can interfere with the interpretation of data obtained from postmortem tissue samples and that appropriately trained personnel can perform decapitation humanely in rodents without anesthesia.

REFERENCES

ACLAD (American Committee on Laboratory Animal Disease). 1991. Detection methods for the identification of rodent viral and mycoplasmal infections. Special topic issue, G. Lussier, ed. Lab. Anim. Sci. 41:199-225.

Aguila, H. N., S. P. Pakes, W. C. Lai, and Y. S. Lu. 1988. The effects of transportation stress on splenic natural killer cell activity in C57BL/6J mice. Lab. Anim. Sci. 38(2):148-151.

Allred, J. B., and G. G. Berntson. 1986. Is euthanasia of rats by decapitation inhumane? J. Nutr. 116:1859-1861.

Amyx, H. L. 1987. Control of animal pain and distress in antibody production and infectious disease studies. J. Am. Vet. Med. Assoc. 191:1287-1289.

AVMA (American Veterinary Medical Association). 1993. 1993 Report of the AVMA Panel on Euthanasia. J. Am. Vet. Med. Assoc. 202:229-249.

Baker, H. J., J. R. Lindsey, and S. H. Weisbroth, eds. 1979. The Laboratory Rat. Vol. I: Biology and Diseases. New York: Academic Press. 435 pp.

Barthold, S. W. 1986. Mouse hepatitis virus. Biology and epizootiology. Pp. 571-601 in Viral and Mycoplasmal Infections of Laboratory Rodents: effects on Biomedical Research, P. N. Bhatt, R. O. Jacoby, H. C. Morse III, and A. E. New, eds. New York: Academic Press.

Beamer, T. C. 1972-1973. Pathological changes associated with ovarian transplantation. P. 104 in The 44th Annual Report of the Jackson Laboratory. Bar Harbor, Maine: The Jackson Laboratory.

Berg, R. J., and C. E. Blass. 1985. Sterilization. Pp. 261-265 in Textbook of Small Animal Surgery, D. H. Slatter, ed. Philadelphia: W. B. Saunders.

Bhathena, S. J. 1992. Comparison of effects of decapitation and anesthesia on metabolic and hormonal parameters in Spraque-Dawley rats. Life Sci. 50:1649-1655.

Bhatt, P. N., R. O. Jacoby, H. C. Morse III, and A. E. New, eds. 1986. Viral and Mycoplasmal Infections of Laboratory Rodents. Effects on Biomedical Research. New York: Academic Press. 844 pp.

Bhatt, P. N., and R. O. Jacoby. 1987. Mousepox in inbred mice innately resistant or susceptible to lethal infection with ectromelia virus. III. Experimental transmission of infection and derivation of virus-free progeny from previously infected dams. Lab. Anim. Sci. 37:23-27.

Bishop, D. H. L. 1990. Arenaviridae and their replication. Pp. 1231-1243 in Virology, B. N. Fields and D. M. Knipe, eds. New York: Raven Press.

Blass, E. M., and E. Fitzgerald. 1988. Milk-induced analgesia and comforting in 10-day-old rats: pioid mediation. Pharmacol. Biochem. Behav. 29:9-13.

Blum, J. R. 1988. Laboratory Animal Anesthesia. Pp. 329-341 in Experimental Surgery and Physiology: Induced Animal Models of Human Disease, M. M. Swindle and R. J. Adams, eds. Baltimore: Williams & Wilkins.

Bradfield, J. F., T. R. Schachtman, R. M. McLaughlin, and E. K. Steffen. 1992. Behavioral and physiologic effects of inapparent wound infection in rats. Lab. Anim. Sci. 42:572-578.

Brown, M. J. 1994. Aseptic surgery for rodents. Pp. 67-72 in Rodents and rabbits: current research issues. S. M. Niemi, J. S. Venable, and H. N. Guttman, eds. Bethesda, Md.: Scientists Center for Animal Welfare. Available from Scientists Center for Animal Welfare, Golden Triangle Building One, 7833 Walker Drive, Suite 340, Greenbelt, MD 20770.

Butler, M. M., S. M. Griffey, F. J. Clubb, Jr., L. W. Gerrity, and W. B. Campbell. 1990. The effect of euthanasia technique on vascular arachidonic acid metabolism and vascular and intestinal smooth muscle contractitility. Lab. Anim. Sci. 40(3):277-283.

Cassell, G. H., J. K. Davis, J. W. Simecka, J. R. Lindsey, N. R. Cox, S. Ross, and M. Fallon. 1986. Mycoplasmal infections: Disease pathogenesis, implications for biomedical research, and control. Pp. 87-130 in Viral and Mycoplasmal Infections of Laboratory Rodents: Effects on Biomedical Research, P. N. Bhatt., R. O. Jacoby, H. C. Morse III, and A. E. New, eds. New York: Academic Press.

CCAC (Canadian Council on Animal Care). 1980. Guide to the Care and Use of Experimental Animals, vol. 1. Ontario, Canada: Canadian Council on Animal Care. 120 pp.

CDC (Centers for Disease Control and Prevention). 1993. Update: Hantavirus pulmonary syndrome—United States, 1993. MMWR 42:816-820.

Clifford, D. H. 1984. Preanesthesia, anesthesia, analgesia, and euthanasia. Pp. 527-562 in Laboratory Animal Medicine, J. G. Fox, B. J. Cohen, and F. M. Loew, eds. Orlando, Fla.: Academic Press.

Cunliffe-Beamer, T. L. 1983. Biomethodology and surgical techniques. Pp. 419-420 in The Mouse in Biomedical Research. Vol. III: Normative Biology, Immunology and Husbandry, H. L. Foster, J. D. Small, and J. G. Fox, eds. New York: Academic Press.

Cunliffe-Beamer, T. L. 1990. Surgical techniques. Pp. 80-85 in Guidelines for the Well-Being of Rodents in Research, H. N. Guttman, ed. Bethesda, Md.: Scientists Center for Animal Welfare. Available from Scientists Center for Animal Welfare, Golden Triangle Building One, 7833 Walker Drive, Suite 340, Greenbelt, MD 20770.

Cunliffe-Beamer, T. L. 1993. Applying principles of aseptic surgery to rodents. AWIC Newsletter 4(2):3-6. Available from the Animal Welfare Information Center, National Agricultural Library, Room 205, National Agricultural Library, Beltsville, MD 20705.

Cunliffe-Beamer, T. L., and E. P. Les. 1987. The laboratory mouse. Pp 275-308 in The UFAW Handbook on The Care and Management of Laboratory Animals, 6th ed., T. Poole, ed. Essex, England: Longman Scientific & Technical.

Derr, R. F. 1991. Pain perception in decapitated rat brain. Life Sci. 49(19):1399-1402.

Dymsza, H. A., S. A. Miller, J. F. Maloney, and H. L. Foster. 1963. Equilibration of the laboratory rat following exposure to shipping stresses. Lab. Anim. Care. 13:60-65.

Engquist, A., M. R. Brandt, A. Fernandes, and H. Kehlet. 1977. The blocking effect of epidural analgesia on the adrenocortical and hyperglycemic responses to surgery. Acta Anaesthesiol Scand. 21:330-335.

Farrar, P. L., J. E. Wagner, and N. Kagiyama. 1986. Syphacia spp. Pp. III.B.1.-III.B.4 in Manual of Microbiologic Monitoring of Laboratory Animals, A. M. Allen and T. Nomura, eds. NIH Pub. No. 86-2498. Washington, D.C.: U.S. Department of Health and Human Services.

Fekety, R., J. Silva, R. Toshniwal, M. Allo, J. Armstrong, R. Browne, J. Ebright, and G. Rifkin. 1979. Antibiotic-associated colitis: Effects of antibiotics on *Clostridium difficile* and the disease in hamsters. Rev. Infect. Dis. 1:386-397.

Feldman, D. B., and B. N. Gupta. 1976. Histopathologic changes in laboratory animals resulting from various methods of euthanasia. Lab. Anim. Sci. 26: 218-221.

Findon, G., and T. E. Miller. 1987. Treatment of *Trichosomoides crassicauda* in laboratory rats using Ivermectin. Lab. Anim. Sci. 37:496-499.

Flecknell, P. A. 1984. The relief of pain in laboratory animals. Lab. Anim. (London) 18(2):147-160.

Flecknell, P. A. 1987. Laboratory Animal Anesthesia: An Introduction for Research Workers and Technicians London: Academic Press. 156 pp.

Flecknell, P. A., and J. H. Liles. 1992. Evaluation of locomotor activity and food and water consumption as a method of assessing postoperative pain in rodents. Pp. 482-488 in Animal Pain, C. E. Short and A. Van Poznak, eds. London: Churchill Livingstone.

Flynn, R. J. 1973. Parasites of Laboratory Animals. Ames: Iowa State University Press. 884 pp.

Flynn, B. M., P. A. Brown, J. M. Eckstein, and D. Strong. 1989. Treatment of *Syphacia obvelata* in mice using Ivermectin. Lab. Anim. Sci. 39:461-463.

Foster, H. L., J. D. Small, and J. G. Fox, eds. 1982. The Mouse in Biomedical Research. Vol. II: Diseases. New York: Academic Press. 449 pp.

Fujiwara, K., and J. E. Wagner. 1986. Sendai virus. Pp. I.G.1-I.G.3, in Manual of Microbiologic Monitoring of Laboratory Animals, A. M. Allen and T. Nomura, eds. NIH Pub. No. 86-2498. Washington, D.C.: U.S. Department of Health and Human Services.

Ganaway, J. R. 1980. Effect of heat and selected chemical disinfectants upon infectivity of spores of *Bacilus piliformis* (Tyzzer's disease). Lab. Anim. Sci. 30:192-196.

Ganaway, J. R., A. M. Allen, T. D. Moore, and H. J. Bohner. 1973. Natural infection of germfree rats with *Mycoplasma pulmonis*. J. Infect. Dis. 127:529-537.

Green, C. J. 1982. Animal Anaesthesia. Laboratory Animal Handbook 8. London: Laboratory Animals Ltd.

Green, S. A. 1991. Pain and analgesia in the post-operative arena. Pp. 589-591 in Proceedings of the 1991 ACVS Veterinary Symposium. San Francisco, Calif: American College of Veterinary Surgeons.

Gurevich, I. 1991. Infection control: Applying theory to clinical practice. Pp. 655-662 in Disinfection, Sterilization and Preservation, 4th ed., S. S. Block, ed. Philadelphia: Lea & Febiger.

Hamm, T. E., ed. 1986. Complications of Viral and Mycoplasmal Infections in Rodents to Toxicology Research and Testing. Washington, D.C.: Hemisphere Publishing Corp. 191 pp.

Hardie, E. M. 1991. Postoperative pain control. Pp. 598-600 in Proceedings of the 1991 ACVS Veterinary Symposium. San Francisco, Calif.: American College of Veterinary Surgeons.

Harkness, J. E., and J. E. Wagner. 1989. The Biology and Medicine of Rabbits and Rodents, 3rd ed. Philadelphia: Lea & Febiger. 230 pp.

Harvey, R. C., and J. Walberg. 1987. Special considerations for anesthesia and analgesia in research animals. Pp. 380-392 in Principles and Practice of Veterinary Anesthesia, C. E. Short, ed. Baltimore: Williams & Wilkins.

Heavner, J. E. 1992. Pain recognition during experimentation and tailoring anesthetic and analgesic administration to the experiment. Pp. 509-513 in Animal Pain, C. E. Short and A. Van Poznak, eds. London: Churchill Livingstone.

Howard, H. L., E. McLaughlin-Taylor, and R. L. Hill. 1990. The effect of mouse euthanasia technique on subsequent lymphocyte proliferation and cell mediated lympholysis assays. Lab. Anim. Sci. 40(5):510 -514.

Hsu, C. K. 1979. Parasitic diseases. Pp. 307-331 in The Laboratory Rat. Vol. I: Biology and Diseases, H. J. Baker, J. R. Lindsey, and S. H. Weisbroth, eds. New York: Academic Press.

Hsu, C. K. 1982. Protozoa. Pp. 359-372 in The Mouse in Biomedical Research. Vol. II: Diseases, H. L. Foster, J. D. Small, J. G. Fox, eds. New York: Academic Press.

Hughes, H. C. 1981. Anesthesia of laboratory animals. Lab. Anim. 10(3):40-56.

Hughes, H. C., W. J. White, and C. M. Lang. 1975. Guidelines for the use of tranquilizers and anesthetics and analgesics in laboratory animals. Vet. Anesth. 2:L19-24.

IATA (International Air Transport Association), IATA Live Animal Regulations. 1995. Montreal, Quebec: International Air Transport Association (IATA). Available in English, French, or Spanish from IATA, 2000 Peel Street, Montreal, Quebec H3A 2R4, Canada (phone: 514-844-6311).

Jenkins, W. L. 1987. Pharmacologic aspects of analgesic drugs in animals: an overview. J. Am. Vet. Med. Assoc. 191(10):1231-1240.

Kanarek, R. B., E. S. White, M. T. Biegen, and R. Marks-Kaufman. 1991. Dietary influences on morphine-induced analgesia in rats. Pharmacol. Biochem. Behav. 38:681-684.

Kehlet, H. 1989. Surgical stress: the role of pain and analgesia. Br. J. Anaesthiol. 63:189-195.

Knecht, C. D., A. R. Allen, D. J. Williams, and J.H. Johnson. 1981. Fundamental Techniques in Veterinary Surgery, 2d ed. Philadelphia: W. B. Saunders. 305 pp.

Kruckenburg, S. M. 1979. Appendix 2: Drugs and dosages. Pp. 259-267 in The Laboratory Rat. Vol. II: Research Applications, H. J. Baker, J. R. Lindsey, and S. H. Weisbroth, eds. New York: Academic Press.

Landi, M. S., J. W. Kreider, C. M. Lang, and L. P. Bullock. 1982. Effects of shipping on the immune function in mice. Am. J. Vet. Res. 43(9):1654 -1657.

LeDuc, J. W., K. M. Johnson, and J. Kawamata. 1986. Hantaan and related viruses. Pp.

I.B.1-I.B.3 in Manual of Microbiologic Monitoring in Laboratory Animals, A. M. Allen and T. Nomura, eds. NIH Pub. No. 86-2498. Washington, D.C.: U. S. Department of Health and Human Services.

Lumb, W. V., and E. W. Jones. 1984. Veterinary Anesthesia. Philadelphia: Lea & Febiger. 693 pp.

Matsushita, S., H. Joshima, T. Matsumoto, and K. Fukutsu. 1989. Transmission experiments of cilia-associated respiratory bacillus in mice, rabbits, and guinea pigs. Lab. Anim. (London) 23:96-102.

McCurin, D. M., and R. L. Jones. 1985. Principles of Surgical Asepsis. Pp. 250-261 in Textbook of Small Animal Surgery, D. H. Slatter, ed. Philadelphia: W. B. Saunders.

McKisic, M. D., D. W. Lancki, G. Otto, P. Padrid, S. Snook, D. C. Cronin II, P. D. Lohmar, T. Wong, and F. W. Fitch. 1993. Identification and propagation of a putative immunosuppressive orphan parvovirus in cloned T cells. J. Immunol. 150:419-428.

Montgomery, C. A., Jr. 1987. Control of animal pain and distress in cancer and toxicological research. J. Am. Vet. Med. Assoc. 191(10):1277-1281.

New, A. E. 1981. Ectromelia (mousepox) in the United States. Proceedings of a seminar presented at the 31st Annual Meeting of the American Association for Laboratory Animal Science. Lab. Anim. Sci. 31(part II):549-635.

NRC (National Research Council), Institute of Laboratory Animal Resources, Committee on Long-Term Holding of Laboratory Rodents. 1976. Long-term holding of laboratory rodents. ILAR News 19(4):L1-L25.

NRC (National Research Council), Institute of Laboratory Animal Resources, Committee on Infectious Diseases of Mice and Rats. 1991a. Infectious Diseases of Mice and Rats. Washington, D.C.: National Academy Press. 397 pp.

NRC (National Research Council), Institute of Laboratory Animal Resources, Committee on Educational Programs in Laboratory Animal Science. 1991b. Education and Training in the Care and Use of Laboratory Animals: A Guide for Developing Institutional Programs. Washington, D.C.: National Academy Press. 139 pp.

NRC (National Research Council), Institute of Laboratory Animal Resources, Committee on Pain and Distress in Laboratory Animals. 1992. Recognition and Alleviation of Pain and Distress in Laboratory Animals. Washington, D.C.: National Academy Press. 137 pp.

NRC (National Research Council), Institute of Laboratory Animal Resources, Committee to Revise the Guide for the Care and Use of Laboratory Animals. 1996. Guide for the Care and Use of Laboratory Animals, 7th edition. Washington, D.C.: National Academy Press.

Oldstone, M. B. A. 1987. The arenaviruses—An introduction. Pp. 1-4 in Arenaviruses, Genes, Proteins, and Expression, M. B. A. Oldstone, ed. Curr. Topics Microbiol. Immunol., Vol. 133. Heidelberg: Springer-Verlag.

Oldstone, M. B. A., and F. J. Dixon. 1967. Lymphocytic choriomeningitis: production of antibody by "tolerant" infected mice. Science 158:1193-1195.

Oldstone, M. B. A., and F. J. Dixon. 1969. Pathogenesis of chronic disease associated with persistent lymphocytic choriomeningitis viral infection. I. Relationship of antibody production to disease in neonatally infected mice. J. Exp. Med. 129:483-505.

OTA (Office of Technology Assessment). 1986. Alternatives to Animal Use in Research, Testing, and Education. Pub. No. OTA-BA-273. Washington, D.C.: U.S. Congress.

Orcutt, R. P., and P. N. Bhatt. 1986. Rat parvovirus. Pp. I.F.1-1.F.3 in Manual of Microbiologic Monitoring of Laboratory Animals, A. M. Allen and T. Nomura, eds. NIH Pub. No. 86-2498. Washington D.C.: U.S. Department of Health and Human Services.

Parker, J. C. 1986. Lymphocytic choriomeningitis. Pp. I.C.1-I.C.5 in Manual of Microbiologic Monitoring of Laboratory Animals, A. M. Allen and T. Nomura, eds. NIH Pub. No. 86-2498. Washington, D.C.: U.S. Department of Health and Human Services.

Parker, J. C., J. R. Ganaway, and C. Gillette. 1987. Viral diseases. Pp. 95-110 in Laboratory

Hamsters, G. L. Van Hoosier, Jr. and C. W. McPherson, eds. Orlando, Fla.: Academic Press.

Phifer, C. B., and L. M. Terry. 1986. Use of hypothermia for general anesthesia in preweanling rodents. Physiol. Behav. 38:887-890.

PHS (Public Health Service). 1996. Public Health Service Policy on Humane Care and Use of Laboratory Animals. Washington, D.C.: U.S. Department of Health and Human Services. 16 pp. Available form the Office for Protection from Research Risks, National Institutes of Health, 6100 Executive Builevard, MSC 7507, Suite 3B01, Rockville, MD 20892-7507

Pick, C. G., J. Cheng, D. Paul, and G. W. Pasternak. 1991. Genetic influences in opioid analgesic sensitivity in mice. Brain Res. 566:295-298.

Powers, D. L. 1985. Preparation of the surgical patient. Pp. 279-285 in Textbook of Small Animal Surgery, D. H. Slatter, ed. Philadelphia: W. B. Saunders.

Rawls, W. E., M. A. Chan, and S. R. Gee. 1981. Mechanisms of persistence in arenavirus infections: A brief review. Can. J. Microbiol. 27:568-574.

Ronald, N. C., and J. E. Wagner. 1976. The arthropod parasites of the genus *Cavia*. Pp. 201-209 in The Biology of the Guinea Pig, J. E. Wagner and P. J. Manning, eds. New York: Academic Press.

Sanford, J. 1992. Guidelines for detection and assessment of pain and distress in experimental animals. Pp. 515-524 in Animal Pain, C. E. Short and A. Van Poznack, eds. London: Churchill Livingstone.

Selwyn, M. R., and W. R. Shek. 1994. Sample sizes and frequency of testing for health monitoring in barrier rooms and isolators. Contemp. Top. 33:56-60.

Selye, H. 1946. The general adaptation syndrome and the diseases of adaptation. J. Clin. Endocrinol. 6:118-127.

Shek, W. R. 1994. Lymphocytic choriomeningitis virus. Pp. 35-42 in Manual of Microbiologic Monitoring in Laboratory Animals, K. Waggie, N. Kagiyama, A. M. Allen, and T. Nomura, eds. NIH Pub. No. 94-2498. Washington, D.C.: U.S. Department of Health and Human Services.

Silverman, J., H. Blatt, and A. Lerro. 1983. Effect of Ivermectin against *Myobia musculi*. Lab. Anim. Sci. 33:487 (abstr).

Skinner, H. H., and E. H. Knight. 1979. The potential role of Syrian hamsters and other small animals as reservoirs of lymphocytic choriomeningitis virus. J. Small Anim. Pract. 20:145-161.

Small, J. D. 1968. Fatal enterocolitis in hamsters given lincomycin hydrochloride. Lab. Anim. Care 18:411-420.

Small, J. D. 1987. Drugs used in hamsters with a review of antibiotic-associated colitis. Pp. 179-199 in Laboratory Hamsters, G. L. Van Hoosier, Jr. and C. W. McPherson, eds. Orlando, Fla.: Academic Press.

Smith, A. L., F. X. Paturzo, E. P. Garnder, S. Morgenstern, G. Cameron, and H. Wadley. 1984. Two epizootics of lymphocytic choriomeningitis virus occurring in laboratory mice despite intensive monitoring programs. Can. J. Comp. Med. 48:335-337.

Smith, A. L., R. O. Jacoby, E. A. Johnson, F. Paturzo, and P. N. Bhatt. 1993. In vivo studies with an "orphan" parvovirus of mice. Lab. Anim. Sci. 43:175-182.

Soma, L. R. 1983. Anesthetic and analgesic considerations in the experimental animal. Ann. N.Y. Acad. Sci. 406:32-47.

Stickrod, G. 1979. Ketamine/xylazine anesthesia in the pregnant rat. J. Am. Vet. Med. Assoc. 175(9):952-953.

Taylor, D. M. 1992. Eradication of pinworms (*Syphacia obvelata*) from Syrian hamsters in Quarantine. Lab. Anim. Sci. 42:413-414.

Thigpen, J. E., E. H. Lebetkin, M. L. Dawes, H. L. Amyx, and G. F. Caviness et al. 1989. The use of dirty bedding for the detection of murine pathogens in sentinel mice. Lab. Anim. Sci. 39:324-327.

Vanderlip, J. E., and B. A. Gilroy. 1981. Guidelines concerning the choice and use of anesthetics, analgesics and tranquilizers in laboratory animals. San Diego, Calif.: Office of Campus Veterinary Science, University of California. 27 pp.

Van Hoosier, G. L., Jr., and L. R. Robinette. 1976. Viral and chlamydial diseases. Pp. 137-152 in The Biology of the Guinea Pig, J. E. Wagner and P. J. Manning, eds. New York: Academic Press.

Van Hoosier, G. L., Jr., and C. W. McPherson. 1987. Laboratory Hamsters. Orlando, Fla.: Academic Press. 400 pp.

Vetterling, J. M. 1976. Protozoan parasites. Pp. 163-196 in The Biology of the Guinea Pig, J. E. Wagner and P. J. Manning, eds. New York: Academic Press.

Waggie, K., N. Kagiyama, A. M. Allen, and T. Nomura, eds. 1994. Manual of Microbiologic Monitoring of Laboratory Animals, 2nd ed. NIH Pub. No. 94-2498. Washington, D.C.: U.S. Department of Health and Human Services. 226 pp.

Wagner, J. E. 1976. Miscellaneous disease conditions in guinea pigs. Pp. 227-234 in The Biology of the Guinea Pig, J. E. Wagner and P. J. Manning, eds. New York: Academic Press.

Wagner, J. E. 1987. Parasitic diseases. Pp. 135-156 in Laboratory Hamsters, G. L. Van Hoosier, Jr. and C. W. McPherson, eds. Orlando, Fla.: Academic Press.

Wagner, J. E., and P. J. Manning, eds. 1976. The Biology of the Guinea Pig. New York: Academic Press. 317 pp.

Wagner, J. E., P. L. Farrar, and N. Kagiyama. 1986. Spironucleus muris. Pp. III.A.1-III.A.3 in Manual of Microbiological Monitoring of Laboratory Animals, A. M. Allen and T. Nomura, eds. NIH Pub. No. 86-2498. Washington, D.C.: U.S. Department of Health and Human Services.

Wallace, G. D., R. M. Werner, P. L. Golway, D. M. Hernandez, D. W. Alling, and D. A. George. 1981. Epizootiology of an outbreak of mousepox at the National Institutes of Health. Lab. Anim. Sci. 31:609-615.

Waynforth, H. B. 1980. Surgical technique. Pp. 89-123 in Experimental and Surgical Technique in the Rat. London: Academic Press.

Waynforth, H. B. 1987. Standards of surgery for experimental animals. Pp. 311-312 in Laboratory Animals: An Introduction for New Experimenters, A. A. Tuffery, ed. Chichester: Wiley-Interscience.

Weisbroth, S. H. 1982. Arthropods. Pp. 385-402 in The Mouse in Biomedical Research. Vol. II: Diseases, H. L. Foster, J. D. Small, and J. G. Fox, eds. New York: Academic Press.

Wescott, R. B. 1976. Helminth parasites. Pp. 197-200 in The Biology of the Guinea Pig, J. E. Wagner and P. J. Manning, eds. New York: Academic Press.

Wescott, R. B. 1982. Helminths. Pp. 373-384 in The Mouse in Biomedical Research. Vol. II: Diseases, H. L. Foster, J. D. Small, J. G. Fox, eds. New York: Academic Press.

West, W. L., J. C. Schofield, and B. T. Bennett. 1992. Efficacy of the "micro-dot" technique for administering topical 1% ivermectin for the control of pinworms and fur mites in mice. Contemp. Top. 31:7-10.

White, W. J., and K. J. Field. 1987. Anesthesia and surgery of laboratory animals. Vet. Clin. North. Am. Small Anim. Pract. 17(5):989-1017.

Yeager, M. P. 1989. Outcome of pain management. Anest. Clin. of N. Am. 7:241.

7

Facilities

Productive research programs that yield reproducible results depend on laboratory animal-care programs that combine good management and appropriate facilities. Such factors as facility location, design, construction, and maintenance influence the quality of animal care and the efficiency of operation. The general guidelines for planning and operating animal facilities described below provide a framework in which specific designs and procedures can be implemented on the basis of professional judgment. Minimal standards applying to the housing of guinea pigs and hamsters are published in *Animal Welfare Standards* (9 CFR 3.25-3.41). The *Good Laboratory Practice Standards* apply to the housing of animals used for studying substances regulated by the Food and Drug Administration (21 CFR 58) and the Environmental Protection Agency (40 CFR 160, and 40 CFR 792). Reports prepared by the Institute of Laboratory Animal Resources for the National Research Council, such as this one, supplement the more general information contained in the *Guide* (NRC, 1996 et seq.). A series of texts on laboratory animals, sponsored by the American College of Laboratory Animal Medicine, provides specific information about the housing needs of mice, rats, hamsters, and guinea pigs (Baker et al., 1979; Balk and Slater, 1987; Ediger, 1976; Hessler and Moreland, 1984; Lang, 1983; Otis and Foster, 1983; Small, 1983; Wagner and Foster, 1976). The *Handbook of Facilities Planning, Volume 2: Laboratory Animal Facilities* (Ruys, 1991) addresses such topics as facility planning and basic design principles. Finally, articles having to do with facility design, construction, and management can be found in various journals and trade magazines.

LOCATION AND DESIGN

The location and design of an animal facility depend on the scope of institutional research activities, animals to be housed, need for facility flexibility, physical relationship to other functional areas, space availability, and financial constraints. The site and design might further depend on whether the facility is located in space initially constructed for housing animals or in remodeled space.

Careful consideration should be given to the location of an animal facility. Initial construction and subsequent operating costs can be influenced by the following:

- local geologic features;
- accessibility of the site;
- prevailing winds and other climatic conditions;
- availability and adequacy of utility and waste-disposal services;
- adjacent properties and buildings;
- suitability of the site for future expansion or building modification;
- state and local regulations and codes; and
- security needs.

Initial construction and subsequent operating costs of a facility can usually be minimized by placing support, care, and treatment areas adjacent to animal-housing space and on a single floor. If the facility extends into adjacent buildings, consideration should be given to placing the animal space on the same level and connecting it by a covered, climate-controlled passage to facilitate movement of animals and equipment.

Centralization Versus Decentralization

In a *centralized* animal facility, support, care, and treatment areas are adjacent to animal-housing space. The facility usually occupies a single floor or building; if it extends into adjacent buildings, the spaces are contiguous. Research personnel come to the animals. In a *decentralized* facility, areas where animals are housed and used are scattered among rooms, floors, or buildings separated by space that is not dedicated to animal care or support. Animal-housing areas are often adjacent to the laboratories in which the animals are used. In this situation, animal-care personnel come to the animals.

Centralization reduces operating costs of a facility because there is a more efficient flow of animal-care supplies, equipment, and personnel; more efficient use of environmental controls; and less duplication of support services. Centralization reduces the need to transport animals between housing and study sites, thereby minimizing the risk of disease exposure. It might also offer greater security by providing more control over access to the facilities and increasing the ease of monitoring staff and animals. A decentralized facility potentially costs more for initial construction because of requirements for environmental systems and controls for separate sites. Multiple cage washers might also be required. Although duplication increases costs, it does provide backups that can be used if a system or equipment fails at one site. Decentralization can reduce traffic at a single site, thereby facilitating disease- or hazard-control or containment programs. Decentralized facilities are generally more accessible to investigators and might offer a more efficent flow of research supplies, equipment, and personnel.

Functional Areas

In addition to the areas used for actual housing of animals, the *Guide* (NRC, 1996 et seq.) recommends making provisions for the following:

- specialized laboratories or individual areas for such activities as surgery, intensive care, necropsy, radiography, preparation of special diets, experimental manipulation, treatment, and diagnostic laboratory procedures;
- containment facilities or equipment if hazardous biologic, physical, or chemical agents are to be used;
- receiving and storage areas for food, bedding, pharmaceuticals and biologics, and supplies;
- space for the administration, supervision, and direction of the facility;
- showers, sinks, lockers, and toilets for personnel;
- an area separate from animal rooms for eating, drinking, smoking, and applying cosmetics;
- an area for washing and sterilizing equipment and supplies and, depending on the volume of work, machines for washing cages, bottles, glassware, racks, and waste cans; a utility sink; an autoclave for equipment, food, and bedding; and separate areas for holding soiled and clean equipment;
- an area for repairing cages and equipment; and
- an area to store wastes before incineration or removal.

Space Requirements

The total space occupied by an animal facility includes program (net) and nonprogram (gross minus net) space. Program space consists of the space allocated to animal housing and various functional areas. Nonprogram space consists of wall thicknesses, dead space, mechanical chases, corridors, stairwells, and elevators. The ratio of program to nonprogram space for facilities designed to house rodents and rabbits has been estimated to be 1:1, and the ratio of housing to support space about 2:3 (Ruys, 1991).

Many design factors influence those ratios, and they serve only as gross estimates of space allocation during planning of a facility. The animal-facility program space required in research institutions can be estimated more accurately by considering the number of faculty or staff using animals, anticipated animal populations, how the animals will be used, the health status of the animals, whether animals of differing health status will be used, and the dimensions of caging and support equipment.

The size of individual animal-holding rooms should be adequate to accommodate standard equipment, especially caging, and to allow adequate space to service both animals and equipment. Room dimensions also should provide flexibility of use. Rooms of 12 × 20 ft (3.7 × 6.1 m) have been suggested as the most efficient for housing mice, rats, hamsters, guinea pigs, and rabbits (Lang, 1980). However, room size should be based on the needs of the program. For example, preference might be given to smaller rooms or cubicles because they offer more opportunity to isolate animals by health status or use. Every effort should be made to provide the greatest amount of space for caging. Aisle space should be kept at a minimum but should be sufficient to allow cage changing, rack sanitation, and other husbandry manipulations.

Relative Relationships of Space

The relative relationship of animal rooms, support rooms, and administrative space should be such that traffic from contaminated to clean areas is eliminated and the efficiency of movement of personnel, equipment, supplies, and animals is maximized. The location of animal-holding space will be determined to a great extent by the location of cage-sanitation facilities.

Corridors, Vestibules, and Anterooms

Rooms in an animal facility can be arranged along single or multiple corridors. The single-corridor arrangement provides more efficient use of space and can be as much as 20 percent less expensive to construct and also less expensive to operate than a comparable facility with multiple corridors (Graves, 1990). A multiple-corridor arrangement allows unidirectional movement, is less congested, and minimizes the potential for cross contamination of the animals.

Corridors should be wide enough to facilitate the movement of personnel and equipment. Although the *Guide* (NRC, 1996 et seq.) recommends a corridor width of 6-8 ft, single-corridor facilities might require wider corridors to reduce congestion.

Entry and exit airlocks and anterooms provide transitional areas between corridors and animal space. They can serve as sound barriers and should reduce the spread of contaminants and allergens. Although airlocks and anterooms slow movement of personnel, animals, supplies, and equipment by doubling the number of doors that must be passed, this slowing provides additional security. Storage of supplies and equipment in airlocks and anterooms should be limited to that essential to support activities in the adjoining animal rooms.

Interstitial Space

Service crews need access to the HVAC system, water lines, drainpipes, and electric connections. The *Guide* recommends making these utilities accessible through service panels or shafts in corridors outside the animal rooms (NRC, 1996 et seq.). Another option is to use an interstitial floor on which equipment can be checked or repaired without requiring entry into the animal facility.

CONSTRUCTION AND ARCHITECTURAL FINISHES

The *Guide* (NRC, 1996 et seq.) describes construction details and architectural finishes suitable for facilities that house rodents. In general, room surfaces should be moistureproof and free of cracks, unsealed utility penetrations, or imperfect junctions that could harbor vermin or impede cleaning. If rooms will be gas sterilized, they should be sealable. The finishes should be able to withstand scrubbing with detergents and disinfectants. All surfaces should be smooth enough to allow rapid removal of water, but floors should have enough traction to be skid-resistant. Surfaces that might be subjected to movement of equipment should be constructed of material that can withstand such movement. Curbs, guardrails, bumpers, door kickplates, and steel reinforcement of exposed corners help to minimize damage. Exterior windows and skylights are not recommended in animal rooms, because they can contribute to unacceptable variations in temperature and photoperiod.

MONITORING

Within an animal facility, the equipment and systems should be monitored to determine whether they are functioning or conforming to predetermined limits or guidelines necessary for successful operation. Temperature, humidity, airflow, air-pressure gradients, and illumination (intensity and photoperiod) in individual animal rooms should be checked. To be effective, a monitoring program should provide accurate, dependable, and timely results. The data collected should be reviewed by personnel who are trained to interpret the results, and the results should be provided to those who are authorized to take corrective action.

SPECIAL REQUIREMENTS

An animal's health status, genotype, or research use might require that it receive special housing. In addition to conventional animal rooms, various levels of barrier or containment housing or other specialized housing might be required to minimize variations that can modify an animal's response to an experimental regimen.

Barrier housing isolates animals from contamination. The degree of isolation depends on the equipment and procedures used and the design and construction of the barrier facility. Rodents usually housed in barrier facilities include microbiologically associated (defined-flora) and specific-pathogen-free rodents, severely immunosuppressed rodents, and transgenic rodents.

In a complete barrier system, isolator-maintained animals are introduced through entry ports. Equipment and supplies enter through an autoclave or other sterilization or disinfection system. Personnel enter through a series of locks in which they remove their clothes and shower before donning barrier-room attire. Cage-washing and quarantine space might be included within such a barrier. Partial barriers differ from complete barriers in construction features, equipment, or operating procedures.

Facilities for animals used in projects that involve hazardous biologic, chemical, or physical agents should be designed so that exposure of personnel and other animals is minimized or prevented. *Biosafety in the Laboratory* (NRC, 1989) describes four combinations of practices, safety equipment, and facilities (animal biosafety levels 1-4) recommended for infectious-disease activities in which laboratory animals are used. Conventional facilities that are consistent in design and operation with the standards described in the *Guide* (NRC, 1996 et seq.) also meet the standards for biosafety levels 1 and 2. Levels 3 and 4 require increasing degrees of containment.

Rodents are sensitive to noise and should be housed away from noise sources (see Chapter 5). The *Guide* describes design and construction features that control noise transmission, including double-door airlocks, concrete (rather than metal or plaster) walls, the elimination of windows, and the application of sound-attenuating materials to walls or ceilings (NRC, 1996 et seq.).

SECURITY

Each facility should consider developing a plan for preventing or minimizing the damage or work disruption that can result from a break-in or malicious damage. Procedures adopted should protect animals and personnel from injury and should protect equipment from theft or damage without creating limitations that adversely affect the quality of care or impede legitimate access to the facility. Administrative responsibility for security should be assigned, with the lines of authority clearly delineated. The plan should be reviewed regularly and modified as needed.

The number, design, and location of windows and doors influences the ability of a facility manager to control access. At the most basic level, physical security consists of key locks on doors. Computer-controlled card-access systems offer the ability to control and record entrance and egress; however, the computer network should be properly maintained and should be tamperproof. Closed-circuit television and motion monitors complement the efforts of security guards.

REFERENCES

Baker, H. J., J. R. Lindsey, and S. H. Weisbroth. 1979. Housing to control research variables. Pp. 169-192 in The Laboratory Rat. Vol. I: Biology and Diseases, H. J. Baker, J. R. Lindsey, and S. H. Weisbroth, eds. New York: Academic Press.

Balk, M. W., and G. M. Slater. 1987. Care and management. Pp. 61-67 in Laboratory Hamsters, G. L. Van Hoosier, Jr., and C. W. McPherson, eds. Orlando, Fla.: Academic Press.

Ediger, R. D. 1976. Care and management. Pp. 5-12 in The Biology of the Guinea Pig, J. E. Wagner and P. J. Manning, eds. New York: Academic Press.

Graves, R. G. 1990. Animal facilities: Planning for flexibility. Lab Anim. 19(6):29-50.

Hessler, J. F., and A. F. Moreland. 1984. Design and management of animal facilities. Pp. 505-526 in Laboratory Animal Medicine, J. G. Fox, B. J. Cohen, and F. M. Loew, eds. Orlando, Fla.: Academic Press.

Lang, C. M. 1980. Special design considerations for animals facilities. Pp. 117-127 in Design of Biomedical Research Facilities. Monogr. Ser. 4. Washington, D.C.: Department of Health and Human Services.

Lang, C. M. 1983. Design and management of research facilities for mice. Pp. 37-50 in The Mouse in Biomedical Research. Vol. III: Normative Biology, Immunology, and Husbandry, H. L. Foster, J. D. Small, and J. G. Fox, eds. New York: Academic Press.

NRC (National Research Council), Institute of Laboratory Animal Resources, Committee to Revise the Guide for the Care and Use of Laboratory Animals. 1996. Guide for the Care and Use of Laboratory Animals, 7th edition. Washington, D.C.: National Academy Press.

NRC (National Research Council), Board on Chemical Sciences and Technology, Committee on Hazardous Biological Substances in the Laboratory. 1989. Biosafety in the Laboratory: Prudent Practices for the Handling and Disposal of Infectious Materials. Washington, D.C.: National Academy Press. 222 pp.

Otis, A. P., and H. L. Foster. 1983. Management and design of breeding facilities. Pp. 17-35 in The Mouse in Biomedical Research. Vol. III: Normative Biology, Immunology, and Husbandry, H. L. Foster, J. D. Small, and J. G. Fox, eds. New York: Academic Press.

Ruys, T., ed. 1991. Handbook of Facilities Planning. Vol. 2: Laboratory Animal Facilities. New York: Van Nostrand Reinhold. 422 pp.

Small, J. D. 1983. Environmental and equipment monitoring. Pp. 83-100 in The Mouse in Biomedical Research. Vol. III: Normative Biology, Immunology, and Husbandry, H. L. Foster, J. D. Small, and J. G. Fox, eds. New York: Academic Press.

Wagner, J. E., and H. L. Foster. 1976. Germfree and specific pathogen-free. Pp. 21-30 in The Biology of the Guinea Pig, J. E. Wagner and P. J. Manning, eds. New York: Academic Press.

8

Rodents that Require Special Consideration

Rodents with a wide variety of valuable genetic characteristics are available for use in many kinds of research (Altman and Katz, 1979a, 1979b; Festing, 1993; Festing and Greenhouse, 1992; Hansen et al., 1981; Hedrich and Adams, 1990). Most are easily maintained with the husbandry techniques discussed in Chapter 5. However, some important research models, especially those with deleterious mutations, require special care. Some—such as mice that carry the homozygous mutation *scid* (severe combined immune deficiency), some strains of mice that carry the homozygous mutation *nu* (nude), and rodents exposed to sublethal irradiation—are so severely immunodeficient that contact with infectious agents of even low pathogenicity can cause severe illness and death, and they require isolation for survival (NRC, 1989). Others have specific requirements for the presentation of food and water; for example, food pellets must be placed on the cage floors and longer than normal sipper tubes are necessary for rodents with mutations that cause dwarfing, and soft diets are essential for mice and rats with mutations in which the incisors fail to erupt (Marks, 1987). Many mutants are subfertile or sterile and require special breeding techniques to maintain the mutation.

A detailed description of the unique husbandry and breeding requirements for each model is beyond the scope of this book. Mating strategies for propagating lethal, sterile, or deleterious mutations have been described (Green, 1981). Those wishing to use mutant rodents should discuss with the investigator or company providing the animals whether there are special requirements for the animals' care and breeding. This chapter will address selected research models: immunodeficient rodents, wild rodents, rodents used for studying aging, mouse and rat models for type I (insulin-dependent) diabetes mellitus, and transgenic mice. Those models are relatively commonly used in research, and information on their husbandry is often difficult to find.

IMMUNODEFICIENT RODENTS

Rodents whose immune systems have been altered through spontaneous mutation, transgenic manipulation, or the application of immunosuppressive drugs or other treatments have long been useful models in biomedical research. However, the immunologic deficiencies that make these animals useful as models often render them susceptible to a host of opportunistic and adventitious infectious agents that would produce few or no effects in immunologically competent animals (Powles et al., 1992; Soulez et al., 1991). The recommendations in this report that cover various rodent species generally apply to immunologically compromised rodents, but much more stringent housing conditions are often required to ensure the health of immunodeficient rodents.

Husbandry

In general, the cages or other implements used to house immunodeficient rodents should be capable of being adequately disinfected or sterilized on a regular basis. The housing systems should be capable of eliminating airborne contamination of the animals and should be capable of being manipulated without exposing the animals to microbiologic contamination during experimentation and routine husbandry procedures. In determining housing and husbandry requirements for maintaining immunodeficient rodents, it is important to consider the effects of various opportunistic and adventitious microorganisms on the type of research being conducted. The length of the study and the research goals will influence the attention to detail needed to prevent infection with such organisms. Maintaining animals in an axenic or microbiologically associated (defined-flora) state might involve a level of effort that is too great and techniques that are too complex for most experimental studies.

Plastic Cages with Filter Tops

This housing system consists of a shoebox cage usually constructed of transparent autoclavable plastic and a separate filter top—a plastic cap with a removable filtration surface in the top. The cap and cage fit together snugly but do not necessarily form a perfect seal. A stainless-steel wire-bar top keeps animals from gaining access to the filter top and provides a food hopper and a holder for a water bottle. An opaque cage can be used, but a transparent cage facilitates routine animal observation without the need to open the cage except for feeding and watering, sanitation, and experimentation. Cages and filter tops and all food, water, and bedding used in those cages should be sterilized.

All changing and manipulation of animals should be done in a laminar-flow work station using aseptic technique. Sterile gloves or disinfected forceps should be used to manipulate animals in any individual cage, and all experimental manipulations should be done so as to minimize or eliminate contamination of the animals. The successful maintenance of animals with this housing system depends directly on rigid adherence to aseptic technique in all aspects of animal and cage manipulation. Although the initial purchase cost of this housing system might seem relatively low compared with that of other systems for housing immunodeficient rodents, the requirement for laminar-flow change stations, sterile supplies, and other operating expenses leads to a substantial continuing cost. Moreover, only minimal mechanical safeguards are built into this system, and success depends absolutely on technique.

A major drawback to using plastic cages with filter tops is that there is a low rate of air exchange between the cage and the room. As a result, bedding might have to be changed more frequently to minimize the buildup of toxic wastes and gases and keep relative humidity appropriately low.

Individually Ventilated Plastic Cages with Filter Tops

This housing system uses plastic cages with filter tops that are constructed and maintained like those previously described. However, an air supply has been introduced into each cage with a special coupling device similar in appearance to the fittings used for automatic watering. Air is supplied to a cage under positive pressure and is exhausted through the filter top. Other ventilation options with respect to positive and negative pressure, as well as a separate exhaust, are also available. Usually, the air supplied to these cages is filtered with a high-efficiency particulate air (HEPA) filter. This system has advantages over the nonventilated plastic cages, but its principal disadvantage is the potential for contamination of the fittings that are used to introduce air into the cages. Rigorous attention must be paid to disinfection of these fittings. The efficiency of this system in protecting immunodeficient animals from infectious agents has not been extensively evaluated.

Isolators

Large isolators capable of housing many rodent cages are commercially available. As discussed elsewhere in this report, isolators are ideal for excluding microorganisms in that they rely very little on individual technique for many husbandry procedures or experimental manipulations. Traditionally, they have been used for housing axenic or microbiologically associated animals. Many varieties of isolators are available; the most common are those made of a flexible bag of vinyl or other plastic material, such as polyurethane. Modern isolators are relatively easy to use and provide investigators and animal-care technicians with easy access to the animals. Special precautions are not needed, because all manipulation is done through built-in glove sleeves with attached gloves. All supplies provided to the isolator are sterilized and are introduced through a port; a chemical sterilization and disinfection procedure is used to decontaminate the outside of the items that have been previously sterilized and wrapped with plastic or other materials that can withstand chemical sterilization or disinfection. Air into and out of the isolator is usually highly filtered. As opposed to plastic cages with filter tops, the isolator offers an advantage in health assessment, in that a large number of animals are maintained as a single biologic unit. Isolators made of rigid plastic with a flexible front offer additional advantages, such as integrated racking, individual lighting, lower operating air pressures, and conservation of space.

Recent advances in construction coupled with the availability of vacuum-packed and irradiated supplies have made isolators for housing immunologically compromised animals a cost-competitive alternative to cages with plastic filter tops.

HEPA-Filtered Airflow Systems

These systems have a variety of forms, including modular chambers, hoods, and racks that are designed to hold cages under a positive flow of HEPA-filtered air. In some instances, plastic cages with filter tops have been used in laminar airflow racks that supply a steady stream of HEPA-filtered air across the cage tops to facilitate air diffusion through the filters. The design of such racks usually involves a blower that pushes air across a HEPA filter and then into a large space (or plenum) that contains thousands of small holes. The holes are designed to permit air to be blown across shelves on which cages are placed. Because many cages must fit on the shelves, there is considerable eddying or turbulence of air across the tops of the cages. Once the cages are pulled forward 10-20 cm beyond the lip of a shelf, the air no longer flows laminarly and mixes with room air. Another system consists of a flexible-film enclosure in which HEPA-filtered air is supplied under positive pressure to a standard rack or group of racks containing filter-topped cages. For both systems, all manipulations must be made in a laminar-flow work station using aseptic technique.

Environmental Considerations

Immunodeficient rodents have been successfully maintained at recommended room temperatures for rodents (NRC, 1996 et seq.). Several theoretical considerations suggest that some immunodeficient rodents, specifically those lacking hair or thyroid glands, might require a higher ambient temperature because of hypothyroidism and poorly developed brown adipose tissue, which reduce the capability for nonshivering thermogenesis (Pierpaoli and Besedovsky, 1975; Weihe, 1984). In practice, such temperatures are not necessary and in fact can be detrimental because they tend to create husbandry problems, including increased decomposition of feed and bedding, increased rate of growth of environmental bacteria, and an uncomfortable working environment for animal-care personnel. In addition, because housing of immunocompromised animals generally requires systems that restrict airflow and heat transfer, temperatures in the animal cages tend to be higher than ambient temperature; therefore, increasing the room temperatures is generally not necessary.

Humidity and ventilation should be consistent with recommendations in the *Guide* (NRC, 1996 et seq.). It is important to remember that many of the containment systems result in increased relative humidity and restrict ventilation. Therefore, animal density, bedding-change frequency, and the relative humidity of incoming air should be adjusted to compensate for some of these differences.

Food and Bedding

Food and bedding for immunocompromised animals should be sterilized or pasteurized to eliminate vegetative organisms. Depending on the method of sterilization selected, fortification of feed with vitamins might be required. Steam sterilization can drastically reduce concentrations of some vitamins and can accelerate the decomposition of some vitamins during storage. Other treatments, such as irradiation, result in much less destruction of these nutrients and so might not require the same degree of fortification of feed before or after sterilization. Adequate validation of the sterilization process is essential to ensure that food or bedding does not serve as a source of contamination.

Water

The water supplied to immunodeficient animals must be free of microbiologic contamination. Sterilization of water is the only sure method of eliminating such contamination. Sterilization can be accomplished by heat treatment, zonation, or filtration. All those processes must be adequately controlled and validated. Other water treatments have been advocated for use with immunocompromised animals, including acidification, chlorination, chloramination, and the use of antibiotics and vitamins. The principal purpose of adding treatment materials to water is to reduce bacterial growth and hence the likelihood of cross contamination in case bacteria are introduced into the water supply. The treatments are not without effects, which can include alteration of bacterial flora, alterations in macrophage and lymphocyte function, reduction in water consumption, and exposure to chlorinated hydrocarbons (Fidler, 1977; Hall et al., 1980; Herman et al., 1982; McPherson, 1963; Reed and Jutila, 1972). In general, the use of the treatments is not an adequate substitute for sterilization of water and should be used only as an adjunct.

Health Monitoring

Many immunodeficient rodents are susceptible to a greater range and incidence of diseases caused by microorganisms than are immunocompetent animals. The lack of a completely functioning immune system often results in more dramatic clinical signs and pathologic changes than would be seen in immunocompetent animals. Because some immunodeficient animals often lack the ability to produce antibodies in the presence of microorganisms, serology is often not useful for diagnosis. Screening for such agents might require the use of immunocompetent sentinel animals of the appropriate microbiologic status. Most commonly, soiled bedding is used as a means of exposing sentinel animals to the immunocompromised animals, and a period of 4-6 weeks of exposure is often required before samples can be taken. Sentinels must be housed under the same environmental conditions and microbiologic barriers as the immunocompromised animals. Health monitoring of animals maintained in individual plastic cages with filter tops is complicated by the potential for contamination of individual cages, as opposed to large groups of cages, with a particular microorganism. Because frequent screening of every cage is not economically feasible, statistical schemes for sampling or batching soiled bedding for exposure of sentinel animals is often required. That is less of a problem with the use of isolators in which large numbers of cages are kept in the same microbiologic space.

Purchase of animals from commercial sources or transfer of animals from other institutions entails some risk with respect to immunocompromised animals. Health status can be compromised during packing, transport, unpacking, and housing of animals. It is important to provide adequate quarantine and stabilization time to allow assessment of the health status of these animals before they are used in experimental procedures. Appropriate precautions should be taken to disinfect the outside of transport containers and to examine them for integrity. Specialized containers have been developed for transport of immunocompromised rodents and should be used whenever possible.

WILD RODENTS

A large number of rodent species have been maintained and bred in a laboratory environment. Wild rodents are used in many fields of research, including genetics, reproduction, immunology, aging, and comparative physiology and behavior. Hibernating rodents, such as woodchucks (*Marmota monax*) and 13-lined ground squirrels (*Spermophilus tridecemlineatus*), are used to study control of appetite and food consumption, control of endocrine function, and other physiologic changes associated with hibernation. Woodchucks are also used as models to study viral hepatitis and virus-induced carcinoma of the liver.

Wild rodents can be obtained by trapping or, in a few instances, from investigators who are maintaining them in the laboratory. Trapping is the simplest way to acquire wild rodents. However, a collector's permit is required in most states, and it is also important to confirm that the species to be trapped, as well as other species in the trapping area, are not threatened or endangered. It is best to begin trapping with an experienced mammalogist.

A search of the literature will locate investigators who maintain feral rodents in a laboratory environment; however, these scientists usually do not maintain enough animals to permit distribution of more than a few. Colonies of wild rodents are listed in the *International Index of Laboratory Animals* (Festing, 1993), in *Annotated Bibliography on Uncommonly Used Laboratory Animals: Mammals* (Fine et al., 1986), and in the Institute of Laboratory Resources (ILAR) Animal Models and Genetics Stocks Data Base (contact: ILAR, 2101 Constitution Avenue, Washington, DC 20418; telephone, 1-202-334-2590; fax, 1-202-334-1687; URL: http://www2.nas.edu/ilarhome/). Several species of the genera *Mus* and *Peromyscus* are more widely used and are available from laboratory-bred sources.

Hazards

Wild-trapped rodents commonly carry pathogens and parasites that are usually not found in or have been eliminated from animal facilities; therefore, appropriate precautions must be taken to prevent disease transmission between feral and laboratory stocks (see Chapter 6). The primary hazard to personnel is getting bitten. Personnel should always wear protective gloves when handling wild rodents. Mice can be handled with cotton gloves (Dewsbury, 1984) or can be moved from place to place in a tall, thin bottle (Sage, 1981). Metal meat-cutter's gloves can be worn under leather gloves for handling larger, more powerful species, such as black rats (*Rattus rattus*) (Dewsbury, 1984). Elbow-length protection, such as leather gloves and gauntlets, should be worn for handling woodchucks because the animals can turn rapidly and bite the inside of the handler's forearm.

Wild rodents can carry zoonotic diseases, such as leptospirosis and lymphocytic choriomeningitis, that are not usually encountered in laboratory-bred rodents (Redfern and Rowe, 1976). Personnel should be offered immunization for tetanus, and anyone that is bitten should receive prompt medical attention. Wild-caught mastomys [*Praomys* (*Mastomys*) *natalensis*] cannot be imported into the United States, because it is a host for the arenavirus that causes the highly fatal Lassa fever.

Care and Breeding

Many small species can be housed in standard mouse and rat cages (Boice, 1971; Dewsbury, 1974a); solid-bottom cages with wood shavings or other bedding are preferred (Dewsbury, 1984). Most small wild rodents are much quicker than domesticated rodents and can easily escape if the handler is not careful. It is advisable to open cages inside a larger container, such as a tub or deep box, to avoid escapes (Dewsbury, 1984; Sage, 1981). Most species do well if given ad libitum access to water and standard rodent diets; however, voles do better on rabbit diets (Dewsbury, 1984; Fine et al., 1986). General guidelines for caring for wild rodents have been published (CCAC, 1984; Redfern and Rowe, 1976). Fine et al. (1986) have summarized and provided references for laboratory care and breeding of kangaroo rats (*Dipodomys* spp.); grasshopper mice (*Onychomys* spp.); dwarf, Siberian, or Djungarian hamsters (*Phodopus sungorus*); Chinese hamsters (*Cricetulus barabensis*, also called *C. griseus* or *C. barabensis griseus*); common, black-bellied, or European hamsters (*Cricetus cricetus*); white-tailed rats (*Mystromys albicaudatus*), fat sand rats (*Psammomys obesus*), voles (*Microtus* spp.), four-striped grass mice (*Rhabdomys pumilio*), and degus (*Octodon degus*). Guidelines on laboratory maintenance of hystricomorph (Rowlands and Weir, 1974; Weir, 1967, 1976) and heteromyid (Eisenberg, 1976) rodents have been published. Mammalogists and other investigators experienced in working with specific species are also excellent sources of information.

Breeding of many wild species is similar to that of domesticated rodents. Some (e.g., voles and deer mice) breed almost as well in captivity as do domesticated species (Dewsbury, 1984). Others (e.g., four-striped grass mice) require special conditions (Dewsbury, 1974b; Dewsbury and Dawson, 1979). A few investigators have reported that breeding of wild *Mus* species is difficult unless running wheels are provided; exercise (up to 10-15 miles/day) apparently causes females to come into estrus and begin a normal breeding cycle (Andervont and Dunn, 1962; Schneider, 1946). Others have not had this problem (Sage, 1981). Pheromones are extremely important in the reproduction of some wild rodents; too frequent bedding changes preclude successful reproduction. A nesting enclosure might be appropriate and should be constructed of a durable material that is easily sanitized, such as plastic or corrosion-resistant metal. Nesting material might improve neonatal survival.

Peromyscus

Peromyscus maniculatus (the deer mouse) and *P. leucopus* (the white-footed mouse) can be maintained with the same husbandry procedures as laboratory mice. A maximum of seven can be housed in 7 × 10 inch plastic cages. Standard rodent feed and water should be give ad libitum. Rabbit or guinea pig feed should not be used, nor should such supplements as fresh vegetables, raisins, and sunflower seeds. Except for breeding, sexes should be housed separately. *Peromyscus* are reasonably cold-tolerant; the suggested temperature is 22-25°C (71.6-77.0°F), and the ambient temperature should not exceed 33°C (91.4°F).

For breeding, single male-female pairs are formed at the age of about 90 days and remain together throughout life. The estrous cycle is 5 days (Clark, 1984). Females caged alone or with other females will not come into estrus. The average reproductive life of *Peromyscus* is 18-36 months. Females should be checked regularly for pregnancies. Copulatory plugs are not a reliable indication of mating, because they are inconspicuous. Lighting is very important in breeding. A 16:8-hour light:dark ratio is generally satisfactory. Continuous light will produce anestrus, and breeding difficulties can sometimes be overcome by reducing the light cycle to a light:dark ratio of 12:12 hours and gradually increasing it to 16:8 over a 3-week period (W. D. Dawson, Peromyscus Stock Center, unpublished). Introduction of a strange male into a cage with a pregnant female can block the pregnancy (Bronson and Eleftheriou, 1963). Gestation is 22 days, except in lactating females, in which it is delayed by 4-5 days. Females enter postpartum estrus about 12 hours after delivery and then remate; therefore, serial litters are produced at 26- to 27-day intervals. Litter size is usually three to six and rarely exceeds eight. Males provide some of the care for the young. Additional information on the care and breeding of *Peromyscus* can be obtained from the Peromyscus Stock Center, Department of Biology, University of South Carolina, Columbia, SC 29208 (telephone, 803-777-3107; fax, 803-777-4002).

Woodchucks

Woodchucks (*Marmota monax*) have been successfully housed indoors in standard cat, dog, or rabbit cages (Snyder, 1985; Young and Sims, 1979) and outdoors in pens or runs (Albert et al., 1976). Enclosures must be carefully secured because a woodchuck can squeeze through any hole large enough to admit its head (Young and Sims, 1979). Each animal should be provided with a nesting box and nesting material, especially if it is housed under conditions that will induce hibernation, for example, in a cold room or, in a cold climate, outdoors in an unheated enclosure. Very thin woodchucks will not survive hibernation (Young and Sims, 1979). Usually, adult females are housed in small groups, and males are housed individually except during breeding season. However, young males and females can be kept together through their first year (Young and Sims, 1979). Food and water should be made available ad libitum. Water should be provided in heavy porcelain bowls. Standard bottles and sipper tubes are not satisfactory, because the animals grip the tubes in their teeth and shake them until they are dislodged from the bottles (Snyder, 1985; Young and Sims, 1979). Woodchucks do well on commercial rabbit diet (Young and Sims, 1979).

AGING COHORTS

Mice and rats have been favored by mammalian gerontologists as experimental models because of their relatively short and well-defined life spans, small size, comparatively low cost, and the large and growing store of information on their genetics, reproductive biology, physiology, biochemistry, endocrinology, neurobiology, pathology, microbiology, and behavior. However, the term *comparatively low cost* is used advisedly. The true cost in 1994 of producing one 24-month-old rat was approximately $200 and a similarly aged mouse $95; the cost for producing one 36-month-old rat was approximately $350 and a similarly aged mouse $175 (DeWitt Hazzard, National Institute on Aging, National Institutes of Health, Bethesda, Maryland, unpublished). The cost to investigators is slightly more than half that amount because production is subsidized by the National Institute on Aging (NIA). A problem faced by investigators who use aged animals is periodic shortages in older cohorts of some strains.

General Considerations

Strictly speaking, *aging* can refer to all changes in structure and function of an organism from birth to death; however, mammalian gerontologists generally confine their experiments to alterations that occur after the onset of sexual maturity and the transition from the juvenile to the young adult phenotype. In sampling for some measure of aging or accruing pathologic conditions, 6-month-old animals will usually provide a normal baseline, and sampling should be carried out at 6-month intervals. Many investigators consider a 24-month-old rodent to be "old"; however, age-related changes in a number of characteristics are often more pronounced in still older animals.

The mean life span (MnLS) of ad libitum-fed (AL-fed), hybrid strains of specific-pathogen-free (SPF) mice or rats is often around 30 months, whereas that for calorically restricted (CR) animals, depending on the regimen used, can be 30 percent longer (see Figures 8.1 and 8.2). Because caloric restriction retards or eliminates common forms of chronic renal disease and a variety of neoplasms, some gerontologists believe that such nutritional management should be the norm. Comparative changes in AL-fed versus CR rodents are increasingly used to test the validity of putative biologic markers of aging rates.

Survivorship in any colony used for gerontologic research should be determined repeatedly. Survival curves for SPF mice and rats should exhibit a classic "rectangularization pattern," that is, a survival curve should nearly parallel the X axis close to the 100-percent survival level for a prolonged period and then decline sharply as the population nears the species' maximum life span (MxLS), which is defined as the age at which only 10 percent of the animals are surviving. A linear survival curve indicates a problem in the population (e.g., exposure to infectious disease). Patterns of age-related pathology within a colony should be repeatedly evaluated through systematic sampling and necropsy of cohorts of various ages (including histologic examination of the major organs). Any animal euthanatized during the course of a study on aging should be necropsied to determine whether the cause of death, such as a specific lesion or neoplasm, could seriously affect the interpretation of the experimental data. For example, the occurrence of lymphoma involving primarily the spleen of old mice of some strains not only decreases survival, but might cause death before other expected findings can occur; this limits the value of these strains in some studies of age-related immunology. A good deal of information is now available on the pathology of aging cohorts of commonly used laboratory mice and rats (Altman, 1985; Bronson, 1990; Burek, 1978; Myers, 1978; Wolf et al., 1988).

Laboratory Mice

There are obvious advantages to using genetically defined strains for research on aging. Inbred or F1 hybrid strains provide a reproducible gene pool, and so permit a more rigorous evaluation of environmental variables, such as caloric restriction. However, in some circumstances, such as longitudinal studies with markers of aging or searches for longevity-assurance genes, the widest possible allelic variability might be desired. For those purposes, systematically outbred animals might suffice, although in the development of such lines, including so-called Swiss mice, the tendency to select breeding pairs for docility and breeding efficiency has resulted in a loss of genetic heterogeneity. An alternative approach is to develop an 8- or 16-way cross between established inbred lines (van Abeelen et al., 1989).

Recombinant inbred mice can also be useful for aging research because they provide a reassortment of linked parental genes (see Chapter 3). Recombinant congenic strains are of special interest for the analysis of polygenic traits (Démant and Hart, 1986; van Zutphen et al., 1991) because they contain a small fraction of the genome of a genetically defined donor line against a genetic background derived from another genetically well-defined strain. For a discussion of the specific uses and relative values of inbred, congenic, recombinant inbred, and nongenetically defined populations, see Gill (1980).

Eight SPF mouse strains, commonly used for gerontologic studies are available from the NIA: inbred strains A/HeNNia, BALB/cNNia, CBA/CaHNNia, C57BL/6NNia, and DBA/2NNia and hybrid strains BALB/cNNia × C57BL/6NNia F1 (CB6F1), C57BL/6NNia × C3H/NNia F1 (B6C3F1), and C57BL/6NNia × DBA/2NNia F1 (B6D2F1). Crl:SW outbred stock is available commercially. Nude mice have also been suggested for gerontologic research (Masoro, 1990), but they are not available from NIA. By using mouse stocks obtained from NIA for research on aging, an investigator avoids changes in genetic characteristics and phenotypes caused by genetic drift in animals from disparate sources (see Chapter 3). An advantage to using well-studied strains is that historical baseline measures are available for comparison, including characteristic age-related pathologic conditions that might complicate the research (see Hazzard and Soban, 1989, 1991, for bibliographies). Life tables for most mouse strains have been published and are summarized by Abbey (1979), and Masoro (1990) presents accumulated data from several sources (see also Green and Witham, 1991). MnLS and MxLS are required in most cases as background data when choosing a strain. More extensive survival data can be obtained from survival curves like those compiled for the SPF colonies of aging NIA mice maintained at the Division of Veterinary Services, National Center for Toxicological Research (NCTR) in Jefferson, Arkansas. An example of such a curve for B6D2Fl (AL-fed versus CR) is presented in Figure 8.1.

A group of related sublines derived from AKR mice and known as SAM (senescence-accelerated mice) have also been developed. SAM mice display multiple pathologic conditions, have an MnLS of as little as 200 days, and have an MxLS of as little as 290 days. They respond to caloric restriction in the same manner as do other strains of mice (Takeda et al., 1981; Umezawa et al., 1990).

Rats (*Rattus norvegicus*)

Four strains are available from NIA: inbred strains BN/RijNia (Brown Norway) and F344/NNia (Fischer 344) and hybrid strains BN/RijNia × F344/NNia F1 (BNFF1) and F344/NNia × BN/RijNia F1 (FBNF1). Inbred strains BUF/N (Buffalo) and LEW (Lewis) and outbred stocks LE (Long Evans), SD (Sprague Dawley), and WI (Wistar) have also been used in research on

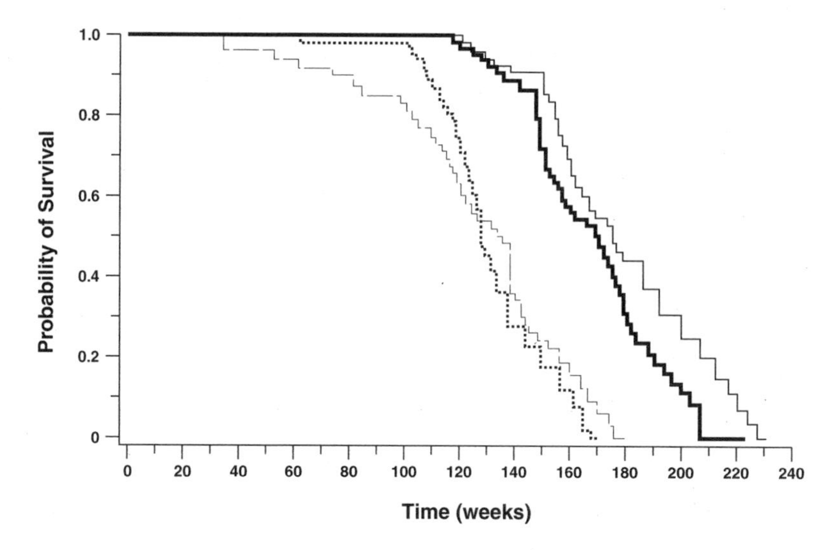

FIGURE 8.1 Survival of male and female C57BL/6NNia × DBA/2NNia F1 (B6D2F1) mice reared under monitored SPF conditions. Studies conducted for the National Institute on Aging by the Division of Veterinary Services, National Center for Toxicological Research, Jefferson, Arkansas. Curves are shown for both AL-fed and CR mice: — —, AL-fed males; . . , AL-fed females; ▬, CR males; ___, CR females. Caloric intake for CR mice was 60 percent of that for AL-fed mice. Calories were reduced gradually between 12 and 16 weeks of age and then continued at reduced levels for the remainder of the life span. All mice were individually housed.

aging. These are available commercially as young animals but seldom as old animals. Life tables are available for each of those stocks and strains (Hoffman, 1979; Masoro, 1990).

Although rats were previously believed to have longer life spans than mice, recent studies indicate that, the life spans of rats and mice are similar (Table 8.1). Rats' larger size might make them more useful than mice for some studies of aging, such as those involving surgery, and rats are widely used in studies on the neurobiology of aging. As do mice, aging cohorts of rats exhibit an increased prevalence of various neoplasms. The prevalence of specific kinds of neoplasms varies among strains. Infectious diseases, including a chronic respiratory complex associated with *Mycoplasma pulmonis*, can also affect life span. The incidence of *M. pulmonis* in rats has been found to be 38 percent in conventionally housed colonies and 0 percent in SPF colonies (NRC, 1991). Thus, cesarean derivation and barrier maintenance can eliminate *M. pulmonis* associated with chronic respiratory disease of rats. Survival curves (AL-fed versus CR) for FBNF1 rats reared under such conditions at NCTR are presented in Figure 8.2.

TABLE 8.1 Mortality for Selected Strains of Mice and Rats Fed Ad Libitum

		Age, weeks			
		Females		Males	
	Strain	50% Mortality	90% Mortality	50% Mortality	90% Mortality
Mice	C57BL/6NNia	117	143	120	141
	DBA/2NNia	77	123	88	126
	C57BL/6NNia × DBA/2NNia F1 (B6D2F1)	128	152	138	171
	C57BL/6NNia × C3H/NNia F1 (B6C3F1)	132	158	140	177
Rats	F344/NNia	116	144	103	121
	BN/RijNia	133	157	129	155
	F344/NNia × BN/RijNia F1	137	166	146	171

SOURCE: Data on National Institute on Aging colonies from the Division of Veterinary Services, National Center for Toxicological Research, Jefferson, Arkansas.

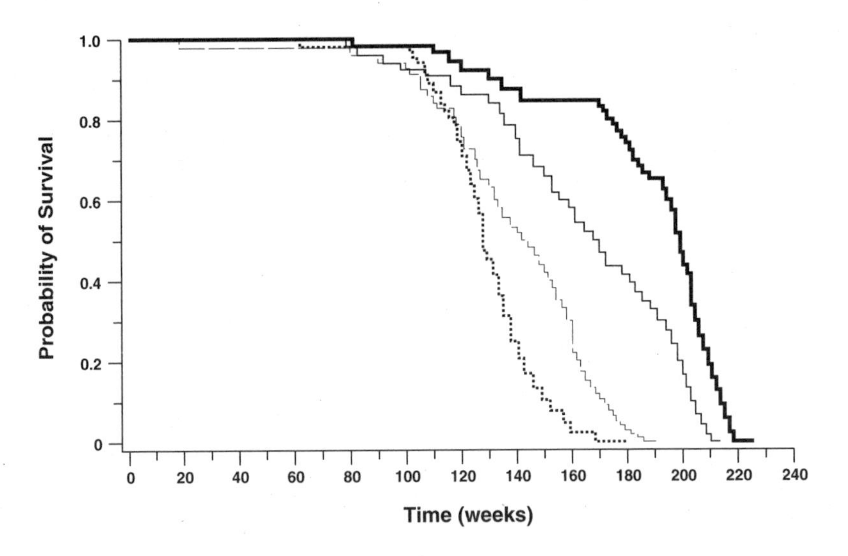

FIGURE 8.2 Survival of male and female F344/NNia × BN/RijNia F1 (FBNF1) rats reared under monitored SPF conditions. Studies conducted for the National Institute on Aging by the Division of Veterinary Services, National Center for Toxicological Research. Curves are shown for both AL-fed and CR rats: — — , AL-fed males; . . , AL-fed females; ▬ , CR males; ——, CR females. Caloric intake for CR rats was 60 percent of that for AL-fed rats. Calories were reduced gradually between 12 and 16 weeks of age and then continued at reduced level for remainder of life span. All rats were individually housed.

Husbandry

There is evidence of an age-related decline in immune response (Miller, 1991), therefore, maintenance of an SPF microbiologic status, under clearly defined and regularly monitored conditions, is a requirement for an aging colony. Mice and rats in an aging colony can be housed in groups (usually four to five animals per cage) or individually. The latter is necessary for both test (CR) and control (AL-fed) animals in caloric-restriction studies. In some colonies, an exercise device, such as a wheel, is provided. The results of studies on whether group housing or exercise facilitation extend MnLS or MxLS vary (Clough, 1991; Holloszy and Schechtman, 1991; Masoro, 1991; Menich and Baron, 1984; Skalicky et al., 1984). A complication of group housing occurs as the old animals begin to die. When that occurs, cages no longer have identical conditions; some contain several animals and others contain only one or two animals. Another complication of group housing, especially among males, is the fighting and threat stress that occurs between animals when dominance is being asserted. The effect of such stress can substantially affect the results of studies on survival, metabolism, and behavior. If males are to be group-housed, they should be grouped immediately after weaning. In some strains, however, this will not prevent fighting. In some instances, the death of one animal in a cage will be followed by the deaths of the rest of the animals in that cage; whether this is caused by an opportunistic pathogen or by the stress of the first animal's death is not clear. Conversely, individual housing is probably stressful initially and might promote inactivity. Thus, the choice of a housing plan depends on the sex and strain of the experimental animals and on the experimental protocol.

Room lighting is especially important in gerontologic research in which performance is measured. Because of the retinal damage that can be caused in albino rodents by exposure to moderately bright light (see Chapter 5), placement of individual cages in relation to the lighting source could influence performance over time. An additional consideration is the light:dark cycle. When CR animals are being compared with AL-fed controls, it is desirable to regulate the light cycle so that both groups will begin eating simultaneously, and activity, cell division, hormone concentrations, and other characteristics will be measured in both groups at similar times on the blood-glucose and -insulin curves. Mice and rats are essentially nocturnal, and AL-fed animals naturally begin feeding shortly after the dark cycle begins. CR animals, in contrast, begin to eat immediately after they are fed, which is usually during the light cycle, and consume most of their food quickly. Both sets of animals can be induced to eat at the same time by reversing the light:dark cycle so that the animal room is dark during the workday. If the light:dark cycle is reversed, the illumination used in the room during the workday should not be visible to the animals.

The temperature of the room and heat-retaining characteristics of the cages are important in studying old or CR animals, which have difficulty in adjusting to cold. Masoro (1991) discusses environmental conditions for aging rats, including the desirability of providing a room temperature somewhat higher than normal. Given the limited knowledge in this regard, a room temperature of 25-27°C (77.0-80.6°F) is suggested for individually housed aging mice and rats, and a somewhat lower temperature for group-housed animals. Variables that will affect this recommendation are the characteristics of the caging (e.g., dispersion of heat through plastic versus through metal and the number of surfaces open to the air) and the airflow and air currents in the room (see Chapter 5).

As discussed previously, diet is a major consideration for aging animals. It affects longevity, perhaps by influencing metabolism and certainly by influencing pathology. Not only caloric restriction, but also the effect of quantity and quality of the protein fed is important (Iwasaki et al., 1988), particularly for strains susceptible to kidney disease. One good high-quality diet is NIH-31, which is used by NCTR for the NIA colonies and by institutions that use animals from the NIA colonies.

Record-Keeping

Record-keeping is discussed in Chapters 4 and 5. Some special considerations apply in aging rodent colonies. In long-term breeding colonies, records of paired-mated sublines should be kept so that selection for life-table characteristics can be either enhanced or limited. Careful records are obviously required for four- or eight-way matings and for the development of recombinant inbred strains. A few animals should be euthanatized and necropsied at regular intervals throughout the study. In the case of mice and rats, this process should begin no later than the age of 18 months.

Transportation and Stabilization

Aged mice and rats are especially susceptible to physical stresses, and this should be a consideration in shipping, as well as in housing the animals. If animals are shipped in very hot or very cold weather, especially if there will be an intermediate holding period in an airport building, they can become debilitated or die. CR mice, in particular, have reduced resistance to cold because of their limited metabolic reserves. It is also difficult to maintain a diet regimen if shipping requires more than 24 hours. The best course of action is to pick up the animals at the airport as soon as they arrive. Transport cartons designed to protect against temperature changes and to maintain SPF status should be used. Arriving shipments of aged SPF rodents should be placed in a barrier facility immediately, even if they will be euthanatized soon after arrival. Failure to do so might lead to bacterial or viral infections that will affect physical performance, immune function, enzyme concentrations, standard blood values, or other characteristics that will be measured. A 2-week quarantine period should be imposed on all arriving shipments of aged animals before they are used in experiments to allow time for incipient infections, if present, to be expressed. Small (1986) has reviewed quarantine periods, particularly with regard to the introduction of communicable diseases (see also Chapter 6). The value of a period to stabilize physiologic and behavioral responses probably varies with the study and should be established by each investigator.

Veterinary Care and Surveillance

Because there is an age-related decline in immune response (Miller, 1991), old mice and rats are especially susceptible to infectious diseases. Therefore, regular microbiologic monitoring (see Chapter 6) is essential for maintaining their SPF status. Sentinel animals should be used for monitoring because aged animals are usually too valuable to euthanatize or to subject to multiple blood-collection procedures. Infectious agents of particular concern to gerontologists are mouse hepatitus virus, Sendai virus, rotavirus, and *Mycoplasma pulmonis* in mice and Sendai virus, Kilham rat virus, rat corona/sialodacryoadenitis virus, and *Mycoplasma pulmonis* in rats (Lindsey, 1986; NRC, 1991). Those agents are of concern because they affect either immune function or general health.

Care of the animals and maintenance of their microbiologic status are usually overseen by the veterinary staff. However, to provide an early warning of incipient health problems, the research staff should observe each animal daily, including weekends and holidays. Moribund or dead animals should be picked up daily before postmortem changes make useful necropsy impossible. A full discussion of barrier facilities and surveillance programs and a summary of infectious disease agents and the systems that they affect have been published (NRC, 1991).

Important considerations to investigators who use aging animals are the timing and method of euthanasia of moribund animals. It is generally considered inhumane to allow old and sick animals to die naturally; however, gerontologic research often requires an accurate record of the time of death. Even if a recorded time of death accurate only to within 24-48 hours would satisfy the experimental protocol, it is difficult to obtain because fragile old mice or rats can appear moribund for days or weeks before they die. Signs of imminent death that can be used to decide when to perform euthanasia are cessation of eating for 48 hours, reduction of body temperature (determined by touching the animals with alcohol-washed fingers or measuring with an electronic thermometer), or maintenance of an immobile posture even if given a gentle stimulus. Each investigator should develop his or her own system with the guidance of the attending veterinarian and, having chosen it, should adhere to it rigorously. An advantage for the investigator of euthanatizing the animal is the ability to obtain usable tissue specimens and necropsy findings. Methods of euthanasia are discussed in Chapter 6.

Other Rodent Species Used for Gerontological Research

Other Species of Mus

A number of interesting species of wild *Mus* and wild subspecies of *Mus musculus* are being adapted for laboratory use (Bonhomme and Guénet, 1989; Potter et al., 1986), but little is known about their life-table characteristics. *Mus caroli* (a rice-field mouse of Southeast Asia) is the single exception. Data on survival, reproductive life span, and age-related pathology have recently been published (Zitnik et al., 1992). The MxLS observed from among cohorts of 249 males and 231 females were 1,560 and 1,568 days, respectively. Gompertz analysis indicated an aging rate only slightly less than that published for wild *Mus musculus*. The shape of the survival curve (especially for females), however, suggests that many animals have died from causes not related to aging, such as fighting and acute stress.

Peromyscus spp.

The best studied member of the genus *Peromyscus* is *Peromyscus leucopus*, the white-footed mouse (Sacher and Hart, 1978), which has a life span about twice that of the laboratory mouse (Sacher, 1977). *Peromyscus*, however, is only "mouse-like"; it has been separated from *Mus musculus* for 15-37 million years. Given that caveat, *Peromyscus* will continue to be useful in broader comparative gerontologic studies because it has adapted well to laboratory conditions. As with all such "domesticated" wild strains, however, a substantial degree of genetic diversity is lost because of the small numbers of animals used to initiate laboratory populations.

Guinea Pigs

The guinea pig (*Cavia porcellus*) has been somewhat neglected by gerontologists because of its comparatively large size, relatively long life span, and relatively high cost of maintenance. Although published survival curves have indicated an MxLS of around 80 months (Rust et al., 1966), some have recorded an MxLS of close to 10 years (Kunst'yr and Naumann, 1984). As with all iteroparous species (species that reproduce more than once in a lifetime) that have not been extensively used for research on aging, the MxLS is likely to be underestimated because record longevities are a function of population size. At least three aspects of guinea pig biology make them of special interest to gerontologists: Like humans, guinea pigs are unable to synthesize ascorbic acid and so are candidates for studies of the free-radical theory of aging (Harman, 1986); their cells appear to be resistant to transformation in vitro (like those of humans and unlike those of mice and rats) (T. H. Norwood and E. M. Bryant, Department of Pathology, University of Washington, Seattle, Washington, unpublished); and the considerable body of research that has been carried out on their auditory system (McCormack and Nutall, 1976) might provide useful background in studies on the pathogenesis of presbycusis.

Guinea pigs are highly susceptible to a variety of infectious diseases; therefore, it is important to maintain them under SPF conditions for gerontologic research. Several such colonies have been established. Husbandry and dietary requirements of guinea pigs have been discussed in Chapter 5.

Hamsters

Primary cultures of Syrian hamster (*Mesocricetus auratus*) somatic cells are often used to study the cellular basis of aging. Cellular function, particularly replicative capacity, can be analyzed in culture with a degree of experimental control that cannot be achieved in living organisms. Normal diploid somatic cells of all studied mammalian species initially divide rapidly in culture, but the replicative capacity or life span of cells is limited, that is it eventually declines. Some of the cells from some species, however, are spontaneously "transformed" and exhibit indefinite replicative potential. Transformation in primary cultures of mouse somatic cells is very rapid and difficult to study, whereas primary cultures of guinea pig somatic cells are resistant to transformation. Syrian hamsters exhibit transformation properties intermediate between those of mice and those of guinea pigs. Investigators interested in a manageable system for studying both the limited replicative life span of cells and their ability escape from such a limitation have found this species to be useful (e.g., Bols et al., 1991; Deamond and Bruce, 1991; Sugawara et al., 1990).

Recent data on survival and pathology are available for a colony of outbred male Syrian hamsters (Deamond et al., 1990). On the basis of 150 spontaneous deaths, the MnLS was 19.5 months, and the MxLS was 36 months. More than 35 inbred strains of Syrian hamsters have been described; most of these have not been carefully investigated in gerontologic research, and many are extinct.

The Turkish hamster (*Mesocricetus brandti*), like other hamsters, offers an opportunity to investigate how hibernation might modify rates of aging and life span (Lyman et al., 1981). The direct correlation found between life span and the amount of time spent in hibernation is consistent with the hypothesis that one or more processes of aging are slowed during hibernation (Lyman et al., 1981).

Chinese hamsters (*Cricetulus griseus*) are of interest to cytogeneticists because their chromosomes are rather easy to study (Brooks et al., 1973). Several outbred, inbred, and mutant stocks have been developed, but they are not as readily available as some other rodents. The life span characteristics of this species have not been rigorously investigated; however, although typical survival curves have been demonstrated for females, the curves for males, which usually live longer, are atypical. An MxLS of about 45-50 months has been reported for males (Benjamin and Brooks, 1977). Information on pathology is available for the colony maintained at the Lovelace Foundation Inhalation Toxicology Research Institute, Albuquerque, New Mexico (Benjamin and Brooks, 1977). Husbandry and dietary requirements have been discussed in Chapter 5.

Gerbils

Cheal (1986) has provided a comprehensive review of the Mongolian gerbil (*Meriones unguicultatus*) as a model for research on aging and has concluded that its ease of handling, ready availability, and particular physiologic and behavioral attributes establish it as a valuable model system. However, the gerbil exhibits an atypical survival curve (Figure 8.3), and much more must be learned about the causes for this, including susceptibility to various infectious diseases and nutritional requirements. All gerbils in the United States are descended from only nine animals (Cheal, 1986), and there is some concern that deleterious recessive or dominant mutations might have become fixed in the population (M. Cheal, University of Dayton Research Institute, Higley, Arizona, unpublished). The husbandry of gerbils is discussed in Chapter 5.

RODENT MODELS OF INSULIN-DEPENDENT DIABETES MELLITUS

With rare exceptions, the rat and mouse models of human autoimmune diabetes mellitus have appeared spontaneously, presumably as a result of mutation, rather than deliberate genetic manipulation. The discussion below focuses on two models of insulin-dependent diabetes mellitus: the BB rat and the NOD mouse. The management principles suggested are easily superimposed on standard rodent-management techniques.

Diabetes-Prone and Diabetes-Resistant Rats

In 1974, some animals were found in a closed colony of outbred WI rats (Bio-Breeding Labs, Ottawa, Ontario) that spontaneously developed autoimmune diabetes mellitus (Chappel and Chappel, 1983). Several inbred diabetes-prone and diabetes-resistant strains were developed from this out-

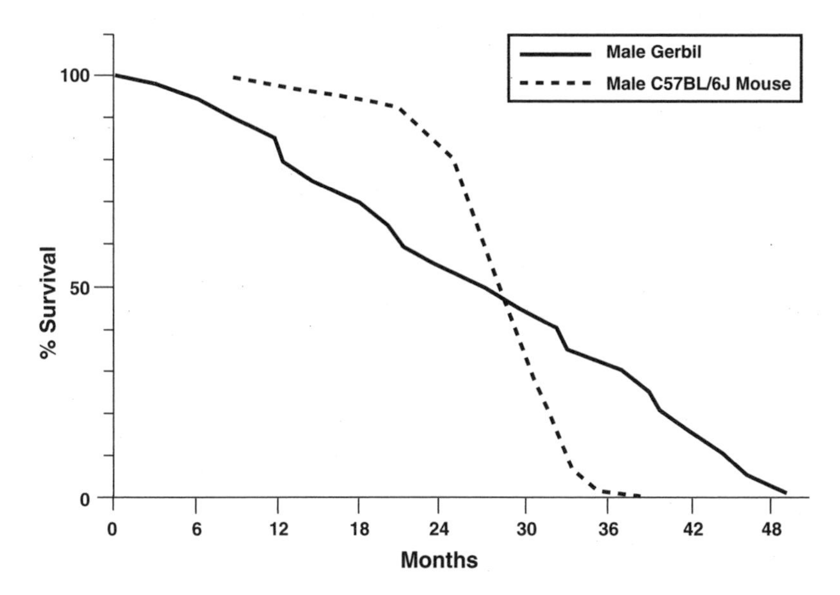

FIGURE 8.3 Survival of conventionally reared male Mongolian gerbils. From Cheal (1986).

bred stock at the Department of Pathology, University of Massachusetts Medical School. The diabetes-prone strains are designated BBBA/Wor, BBDP/Wor, BBBE/Wor, BBNB/Wor, and BBPA/Wor; the diabetes-resistant strains are designated BBDR/Wor and BBVB/Wor.[1] The genetics and pathophysiology of the diabetes-prone strains have been reviewed (Guberski, 1993; NRC, 1989).

Breeding Techniques and Genetic Records

Foundation colonies of diabetes-prone and -resistant strains are maintained strictly by full-sib matings. However, the selection of litters from which future generations of breeders will be derived is influenced by the presence of desired phenotypic traits (e.g., incidence of diabetes, age at onset of diabetes, fertility, litter size, and survival of pups to weaning). Although it is recognized that the imposition of selection criteria can delay achieving inbred status, the goals of any breeding strategy must include preservation of the desired phenotypic characteristics (e.g., the development of diabetes mellitus).

[1]The designation BB/Wor was originally used as a group name for all seven inbred strains.

Essential data on each litter produced in the foundation colonies must be recorded to permit genetic tracing of breeding stock from one generation to another. To achieve this, a system of identification of each member of the primary and secondary breeding branches must be established. The records should include the occurrence of phenotypic characteristics, such as diabetes, thyroiditis, and lymphopenia.

Husbandry and Care

It is desirable that diabetes-prone and -resistant rats be maintained free of rodent pathogens in appropriate barrier facilities (see Chapter 5) because of the effect of these pathogens on phenotypic expression of diabetes (reviewed by Guberski, 1993). Microbiologic status should be monitored and recorded; records should include the tests performed and the frequency of testing. Experience has shown that these animals do well on a conventional light:dark ratio of 12:12 hours.

Detection and treatment of diabetes mellitus. The most cost-effective method of screening for diabetes is to test for glycosuria. Urine is expressed from the bladder manually by gently compressing the bladder against the pubic symphysis. Urinary glucose concentration is measured with a glucose test strip. Positive urine tests are confirmed with blood glucose measurements. Blood samples should be obtained from the tail within 2 hours of the urine test and tested with an appropriate technique. Animals testing 4+ for glycosuria and having blood glucose concentrations greater than 250 mg/dL are considered diabetic.

The age at which to begin testing and the frequency of testing for diabetes depend on the unique characteristics of the particular model and the environmental conditions under which it is kept. Testing for glycosuria should be started before the expected onset of diabetes and performed at least three times per week at the start of the light period in the light-dark cycles. The frequency of glycosuria testing can be reduced after about 120 days because new occurrences are less likely.

Daily treatment of diabetic rats with insulin is mandatory and should begin on the day that glycosuria is found and diabetes is confirmed. The daily dose of insulin will be a function of age, body weight, the presence of ketoacidosis and dehydration, and the presence of pregnancy or lactation. Table 8.2 provides guidelines for the initial doses of insulin for animals that become diabetic after the age of 65 days. Animals that become diabetic *on or before the age of 65 days* should receive 0.2 U of insulin per 100 g of body weight in addition to the dose indicated. As animals increase in weight, the dose of insulin is increased by 0.2 U/10 g of body weight if the animals became diabetic on or before the age of 65 days, and by 0.2 U/16 g of body weight if the animals became diabetic after the age of 65 days. The maximal daily dose should not exceed 1.4 U/100 g of body weight for animals that became diabetic on or before 65 days of age, and 1.25 U/100 g of body weight for animals that became diabetic after the age of 65 days.

If ketonuria (as detected with a test strip) develops, the dose of insulin should be increased, and lactated Ringer's solution with sodium bicarbonate should be administered in the amounts shown in Table 8.3. Injections of fluids are well tolerated when given under the loose skin on the back (distal to the nape of the neck).

Treatment of hypoglycemia. Hypoglycemia is defined as severe if blood glucose is less than 40 mg/dL, moderate if blood glucose is 40-60 mg/dL, and mild if blood glucose is 60-80 mg/dL. The successful treatment of hypoglycemia requires a decrease in insulin dose combined with subcutaneous injections of fluid. Suggested regimens are outlined in Table 8.3.

Care of pregnant females. If pregnant animals become aglycosuric, the course of action depends on the ratio of insulin to "ideal" body weight

TABLE 8.2 Starting Doses of Insulin for BB/Wor Rats That Become Diabetic After the Age of 65 Days

	Initial Blood Glucose Concentration, mg/dL					
	250	300	350	400	450	500+
Body weight, g[a]	Starting Dose of Insulin,[b] U					
100	0.4	0.6	0.6	0.6	0.8	0.8
125	0.4	0.6	0.6	0.8	0.8	0.8
150	0.6	0.8	0.8	1.0	1.0	1.2
175	0.8	1.0	1.0	1.2	1.2	1.4
200	1.0	1.2	1.2	1.4	1.6	1.6
225	1.2	1.4	1.4	1.6	1.6	1.8
250	1.4	1.6	1.6	1.8	1.8	2.0
275	1.4	1.6	1.8	1.8	2.0	2.0
300	1.4	1.6	1.8	2.0	2.0	2.2
325	1.6	1.8	2.0	2.0	2.2	2.2
350	1.6	1.8	2.0	2.2	2.2	2.4
375	1.8	2.0	2.2	2.2	2.4	2.4
400	2.0	2.2	2.4	2.4	2.6	2.6
425	2.2	2.4	2.6	2.6	2.8	3.0
450	2.2	2.4	2.6	2.8	3.0	3.2

[a]Assumes that rat is well hydrated and that ketosis, if present, is being corrected.

[b]PZI U40 (Eli Lilly) insulin and a U/100 Lo-dose syringe (B-D) are used. U40 insulin + U/100 syringe = 0.4 units per gradation mark. Add 0.2 U/100 g of body weight to the dose for animals that develop diabetes on or before the age 65 days. Maximal daily dose equals 1.4 U/100 g of body weight for animals that become diabetic on or before the age of 65 days and 1.25U/100 g of body weight for animals that become diabetic after the age of 65 days.

TABLE 8.3 Treatment for Ketonuria in BB/Wor Rats

Ketones	Increased Insulin,[a] U/100 g body wt	Lactated Ringer's Solution, cm^3	Sodium Bicarbonate, mEq[b]
2+	0.2	10.0	0.0
3+	0.2	9.0	1.0
4+	0.2	18.0	2.0

[a]Insulin dose of lactating females should not exceed 1.0 U/100 g of "ideal" body weight (see Care of pregnant females). Dose should not be increased during mild episodes of ketonuria.
[b]1 cm^3 of 8.4% sodium bicarbonate equals 1 mEq.

SOURCE: Guberski, 1993.

(IBW). The IBW of a pregnant female at the age of 90 days is considered to be 270 g. If the animal is more than 90 days old, the body weight of a nonpregnant female sibling should be used as the IBW. The following procedures are recommended:

- If the ratio of insulin to IBW is greater than 1.0 U/100 g, the dose of insulin should be reduced by 15 percent.
- If the ratio of insulin to IBW is 0.9-1.0 U/100 g, the dose of insulin should be reduced by 10 percent and 10 cm^3 of lactated Ringer's solution should be administered.
- If the ratio of insulin to IBW ratio is less than 0.9 U/100 g, the dose of insulin should be reduced by 0.2 U/100 g and 10 cm^3 lactated Ringers solution should be administered.

If pregnant animals are severely hypoglycemic, follow the instructions for treating hypoglycemia in Table 8.4.

If a female becomes ketotic at parturition, the insulin dose should not be changed. Instead, lactated Ringer's solution and sodium bicarbonate should be injected subcutaneously in the amounts indicated in Table 8.3.

Care of lactating females. Beginning 12-14 days after delivery, insulin should be decreased by 10-15 percent each day until a dose of 0.8-1.0 U/100 g of IBW is achieved. To prevent hypoglycemia in lactating females, food should be made readily accessible by placing it on the cage floors. If hypoglycemia occurs, it should be treated as indicated in Table 8.4.

Use of Spleen Cells to Reduce Frequency of Diabetes and Improve Breeding Efficiency

Diabetes-prone rat strains are profoundly T-cell lymphopenic. Injections of neonatal bone marrow, fresh spleen cells, or concanavalin-A-stimu-

TABLE 8.4 Treatment for Hypoglycemia in Diabetic BB/Wor Rats

Classification (blood glucose concentration)	Subcutaneous Fluid Therapy	Change in Insulin Dose	Change in Time of Insulin Administration
Severe (<40 mg/dL)	Give 1 cm^3 50% dextrose; 2 hrs later give lactated Ringer's solution with 5% dextrose	Reduce by 30-50%	Delay by 2-3 hrs
Moderate (40-60 mg/dL)	Give 10 cm^3 lactated Ringer's solution with 5% dextrose	Reduce by 20-30%	Delay by 2-3 hrs
Mild (60-80 mg/dL)	Give 10 cm^3 lactated Ringer's solution	Reduce by 10-15%	No delay

SOURCE: Guberski, 1993.

lated spleen cells correct the T-cell lymphopenia and substantially reduce the frequency of spontaneous diabetes (Naji et al., 1981; Rossini et al., 1984). Fresh spleen cells are obtained from diabetes-resistant rats, which are histocompatible with diabetes-prone rats but are not lymphopenic. Spleens are prepared with standard techniques (Burstein et al., 1989). Diabetes prone rats between 21 and 40 days old receive one spleen equivalent of fresh donor cells in 1 cm^3 of RPMI medium 1640, administered intraperitoneally. This procedure reduces the incidence of diabetes from greater than 85 percent to about 15 percent. Nondiabetic females do not require daily insulin injections (this reduces the workload of the staff) and are more productive breeders, as shown in Table 8.5.

Shipping Pathogen-Free Rats

Diabetes-prone rats have severely compromised immune systems and should be shipped in crates designed to keep them free of rodent pathogens (see Chapter 6). Drinking water or a water-rich material must be provided, especially for diabetic rats showing signs of polydipsia and polyuria, because these animals are prone to dehydration. Commercial carriers should be instructed to use climate-controlled trucks and holding rooms because diabetic rats are more susceptible than normal rats to fluctuations in temperature. In addition, commercial carriers must guarantee delivery within 24 hours because shipping delays are hazardous for animals that require daily insulin injections.

TABLE 8.5 Reproduction in Diabetes-Prone BB/Wor Rats Before and After Receiving Splenocytes from Diabetes-Resistant BB/Wor Rats

	Diabetes-Prone Females Not Treated with Splenocytes (N = 1,238)	Diabetes-Prone Females Treated with Splenocytes (N = 1,022)
Incidence of diabetes	86%	16%
No. pups born	7,160	12,434
No. pups weaned	5,766	10,918
Pup survival through weaning	80.5%	87.8%
No. pups weaned per female mated	4.7	10.7

SOURCE: Guberski, 1993.

NOD Mice

NOD (nonobese diabetic) is an inbred strain derived from Jcl:ICR mice with selection for the spontaneous development of insulin-dependent diabetes (Makino et al., 1980). The expression of diabetes in this strain is under polygenic control (Leiter, 1993). Clinical features of diabetes in NOD mice are similar to those in humans. Females develop diabetes at a higher incidence and at an earlier age than males. The genetics and pathophysiology of this model have been reviewed (Leiter, 1993; NRC, 1989).

Insulin treatment is required to maintain diabetic NOD mice; without insulin, they survive only 1-2 months after diagnosis. Diabetes is diagnosed by determining that the blood (nonfasting) or plasma glucose concentration is increased. This determination can be made by measuring blood glucose directly or by measuring urinary glucose with a glucose test strip. Glycosuria, as read on the test strip, usually denotes a plasma glucose of 300 mg/dL. Large numbers of mice can be easily screened by this method.

It is difficult to keep serum glucose within a normal range with insulin treatment, but body weight can be maintained and life prolonged (Ohneda et al., 1984). Morning and evening intraperitoneal injections of a 1:1 mixture of regular and NPH insulin are satisfactory. The dose will be 1-3 U, depending on the extent of glycosuria.

Environmental factors are extremely important in the expression of diabetes in NOD mice. Keeping them in an SPF environment increases the occurrence of diabetes; exposure to a variety of murine viruses, including mouse hepatitis virus (Wilberz et al., 1991) and lymphocytic choriomeningitis virus (Oldstone, 1988), prevents diabetes development. That various types of exogenous immunomodulators prevent the development of diabetes (Leiter, 1990) suggests that infectious agents prevent diabetes by general immunostimulation. Diet also has an important effect on diabetes development: natural-ingredient diets, including standard, commercially available mouse feed, promote a high incidence of diabetes (Coleman et al., 1990).

NOD is an inbred strain and should be maintained by brother × sister mating. NOD mice have an excitable disposition but breed well. Siblings bred before the development of overt diabetes can usually produce two large litters (9-14 pups each) of which nearly all the pups survive to weaning. Breeders can be protected from developing diabetes by a single injection of complete Freund's adjuvant (Sadelain et al., 1990).

TRANSGENIC MICE

Since the late 1970s, advances in molecular biology and embryology have enabled scientists to introduce new genetic material experimentally into the germ lines of mice and other animals. The term *transgenic mice*, as used here, means that foreign DNA has been introduced into mice and is transmitted through the germ line. The gene transfer can be performed to introduce new genetic traits or to negate or "knock out" host-gene function by targeted mutagenesis.

Foreign genetic sequences can be introduced into mouse cells, especially in early embryos, by several different methods. The most commonly used method is pronuclear microinjection, in which a solution of purified DNA is injected into either of the two pronuclei visible in a newly fertilized egg (Gordon et al., 1980). Other, less reliable methods include the carrying of the proviral DNA into the cell with a retroviral vector (Jaenisch, 1976) or by electroporation (Toneguzzo et al., 1986) and transformation of totipotent embryonic stem (ES) cells, which are derived from cultured blastocyst-stage embryos (Doetschman et al., 1987). In contrast with microinjection or retroviral insertion, integration of foreign DNA into ES-cell chromosomes can be targeted to specific loci. The specifically modified, undifferentiated ES cells can then be introduced into a recipient embryo in which (it is hoped) they will incorporate into the developing germ line. This approach is used not only for modifying gene expression, but often for introducing targeted mutations by replacement of genes with nonfunctional counterparts, that is, for producing "knockouts" (Mansour et al., 1988).

Colony Management

Although a transgene causes only a small change in a genome, it can produce dramatic and unpredictable changes that make colony maintenance a challenge. Husbandry and production of transgenic mice have been reviewed (Gordon, 1993) and will be described briefly here.

Colony management can be complicated by several characteristics of transgenic mice, including unpredictable phenotypic effects of transgene expression, pathologic effects of the transgene that compromise viability, unpredictable interactions between the transgene and other host genes (e.g., insertional mutagenesis), altered responses to microorganisms or other environmental variables, compromised fertility, and possible instability of transgene expression through generations. Depending on the presence and severity of those characteristics, barrier maintenance might be advisable. Filter-top caging systems are usually sufficient if proper precautions are taken. Flexible-film or rigid isolator systems, however, permit the most complete control of the physical and microbiologic environment. Microbiologic status should be monitored regularly and should include testing for standard murine infectious agents. Both transgenic and sentinel mice should be evaluated if the integration or expression of a foreign gene alters immune competence.

Transgenic mice should be observed daily, and all visible clinical events should be recorded. Animal-care technicians should be trained to recognize clinical events and to report their occurrences with appropriate descriptive terminology. Unexpected deaths should be discussed with an animal-health professional, such as an animal pathologist, to determine whether necropsy and histologic examination are warranted. It is imperative that deceased animals be collected and preserved properly as soon as they are discovered. Corpses can be placed in fixative, refrigerated, or frozen, depending on the specific postmortem procedures that are planned.

Management of a transgenic-mouse facility includes special requirements for embryo donors, embryo recipients, and offspring. In many transgenic facilities, embryo collection and culture, DNA introduction, and embryo transfer are performed outside the barrier; therefore, the embryos and embryo-transfer recipients might no longer be SPF and should not be returned to the barrier.

Embryo Donors

Embryos into which DNA will be introduced to generate founder mice are obtained by administering exogenous gonadotropic hormones intraperitoneally to virgin females. The hormones elicit synchronized ovulation of a relatively large cohort of mature oocytes (i.e., superovulation); therefore, fertilization and later preimplantation development will also be synchronized. Very young females—28-40 days old, depending on the stock or strain—usually respond best to superovulatory hormones. Outbred mice were originally used as embryo donors; more recently, inbred FVB mice have also been used. FVB mice are highly inbred, they respond well to superovulatory hormones, and their embryos have large pronuclei (Taketo et al., 1991).

Males should be individually housed; females can be group-housed before mating. Breeding is most effective if a 3- to 8-month-old male that is a proven breeder is paired and bred with one or two females every 2 or 3 days. Mating should always occur in the cage of the male.

An uninterrupted dark phase of the lighting cycle is critical for efficient superovulatory breeding; a light:dark ratio of 14 to 10 hours is effective. Two gonadotropic hormones, pregnant mare serum gonadotropin (PMSG) and human chorionic gonadotropin (HCG), are each administered 7-9 hours before the beginning of the dark cycle, but PMSG is administered 2 days before HCG. Pronuclear embryos are generally collected 14-17 hours after the beginning of the dark cycle. For example, if the dark cycle begins at 10 p.m., PMSG would be administered between 1 and 3 p.m. 2 days before the day of mating, HCG would be administered between 1 and 3 p.m. on the day of mating, and pronuclear embryos would be collected between noon and 3 p.m. the next day.

Embryo Recipients

Group-housed females are used; outbred or hybrid mice generally make the best dams. Good choices of stocks to carry transferred embryos include outbred ICR mice (if a white coat is desired) and C57BL/6 × DBA/2 F1 (B6D2F1) hybrid mice (if a colored coat is desired). Housing strategies that avoid synchronization of estrus in group-housed females have been described (Gordon, 1993).

A colony of vasectomized males is required. It is preferable for the males to be test mated to ensure sterility; however, if 5- to 6-week-old males are vasectomized, there is no sperm yet in the vas deferens, and test mating is not necessary. Even if test mated, males used to produce pseudopregnant females should be a different color from the embryo donor so that "accidental" offspring of males that have recovered their fertility can be distinguished from transgenic offspring.

Embryo-donor females should be 0-1 day more advanced in the reproductive cycle than pseudopregnant females. Early (one or two cells) embryos are transferred into the oviduct of the embryo recipient; morula and blastocyst embryos are transferred directly into the uterus. Recipient females should be used only once.

Offspring

Individual litters should be separated by sex at weaning and housed in cages that clearly indicate the litter number, date of birth, lineage, and parental identities. In general, fewer than 25 percent of live-born pups that receive transgene DNA as embryos will have integrated transgenes; 10 percent is considered average if microinjection is used. Most transgenic mice are identified by Southern blotting or polymerase chain reaction (PCR) analysis of DNA extracted from tissue taken from the tip of the tail; approximately 1 cm of tissue is sufficient. Rarely, it is possible to identify transgenic mice by detecting gene products from the introduced DNA.

Breeding Transgenic Mice

Once a mouse is identified as transgenic, it should be bred to verify that the transgene has been integrated into its germ cells. The development of a colony of mice homozygous for the transgene is achieved by standard breeding and test-mating procedures. Homozygous transgenic mice will produce 100 percent transgenic progeny on mating with a nontransgenic mate, whereas hemizygotes will produce both transgenic and nontransgenic offspring. It is recommended that multiple test litters be analyzed before the homozygosity of a breeder is considered established. Transgenic inheritance patterns do not always conform to classical Mendelian patterns, because the integration and expression of a transgene can affect implantation, in utero development, and postnatal survival. When mice are not homozygous for the transgene, all offspring must be screened for the transgene.

Reproductive performance of transgenic mice can differ substantially from that of the nontransgenic parental or background strains. Insertional phenomena can compromise fertility and affect embryo survival. Although breeding mice to homozygosity for the transgene is often desirable, homozygotes might be inviable, infertile, or subfertile. If fertility problems are encountered in homozygotes, whether caused by transgene expression or insertional mutagenesis, the problem can often be effectively managed by maintaining the transgene in the hemizygous state. Even in hemizygous mice, however, the effects of transgene integration, transgene expression, or both can be detrimental to survival and reproduction, and investigators and animal-care personnel should be alert to the necessity for establishing aggressive breeding programs. In extreme cases, assisted-reproduction technologies (e.g., superovulation and in vitro fertilization) might be helpful.

Identification, Records, and Genetic Monitoring

Identity, breeding, and pedigree records must be fastidiously kept because breeding errors in transgenic colonies are difficult to detect. For example, classic genetic monitoring will not necessarily distinguish between different transgenic lines on the same background strain. Even direct examination of the transgenic DNA sequence (e.g., with Southern blotting or PCR analysis) might not definitively identify a specific mouse. It is recommended that a combination of methods for identification and genetic monitoring be used in a colony of transgenic mice. Purified DNA samples from important animals can be frozen and stored at –70°C; these might be useful for future analyses, especially if DNA rearrangement is suspected.

Individual animals can be marked rapidly and inexpensively by tattooing, clipping ears, or using ear tags. The most reliable, albeit most expensive, system for identifying an individual animal is subcutaneous implantation of a transponder encoded with data on the animal. Transponder identification chips are durable for the life of the animal and suitable for computerized data-handling. Whatever method is chosen should be used in conjunction with a well-maintained cage-card system. One issue that arises in colonies of genetically engineered animals that does not arise in other colonies is confidentiality specifically related to patentability of the animals; information displayed on cage cards should be reviewed with the principal investigator.

The identity of each transgene-bearing breeder should be verified before mating. Important information on the transgenic parent includes transponder code or other identification code, lineage, date of birth, date of pairing, administration of exogenous hormones (if any), and date of separation of breeding pair. If mice escape, all unidentifiable animals should be euthanatized, and recaptured identifiable females should be isolated for at least 3 weeks to determine whether they are pregnant. Litters derived from questionable or unverified matings should be euthanatized.

Embryo Cryopreservation

Because each transgenic line is unique, embryo cryopreservation might be considered. In general, cryopreservation issues relevant to transgenic lines are the same as those relevant to other rodents (see Chapter 4). However, some lines cannot be made homozygous, are reproductively compromised, or both, so it might be prudent to freeze more embryos than would be necessary for preservation of an inbred strain.

Data Management

A large amount of data accumulates in a transgenic colony and must be managed efficiently. Daily or weekly records include data on breeding, birth, weaning, death, and laboratory analyses; they also include documentation of observations on such things as characteristics that are possibly related to gene manipulation, pathologic conditions, and unusual behaviors.

Shipment and Receipt of Transgenic Rodents

In general, it is not necessary to use extraordinary containment procedures for shipping transgenic mice. To reduce the risk of loss, shipments can be split so that accidents or errors during transit do not compromise the entire shipment. The following information should accompany transgenic mice shipped from a facility and be requested for transgenic mice brought into a facility:

- genetic identity, including the species and strains from which the transgene originated, the designations of all transgene components, the ancestry of the transgenic founder, and the exact lineage designation and generation number of each mouse;
- standardized transgene symbol (see NRC, 1993);
- individual identification numbers accompanied by an explicit description of the identification method (e.g., subcutaneous transponders, 16-digit codes, or an ear-marking scheme with a drawn key);
- description of the predicted phenotype and relationship of transgene expression to such factors as age, sex, pregnancy, and lactation;
- identification of potential human health hazards related to transgene expression (e.g., active expression of intact virus particles or potentially immunogenic viral structural proteins);
- general health status of the mice and probable morbidity or mortality associated with transgene expression, including available data on serologic, bacteriologic, and parasitologic screening; and
- information important to maintenance and breeding, such as breeding strategies, pregnancy rates, gestation times, litter sizes, and sex distribution within litters.

Human Health Hazards

Consideration must be given to possible zoonotic hazards posed by transgenic mice. For example, viral replication has been demonstrated in mice carrying the entire hepatitis B virus genome (Araki et al., 1989). Preliminary banking of employees' sera should be considered (see Chapter 2).

Administrative Issues

In maintaining colonies of transgenic animals, all relevant legal requirements must be addressed. Examples include laws governing patent applications or awards, international regulations governing the importation or exportation of genetically engineered animals, and quarantine laws.

REFERENCES

Abbey, H. 1979. Survival characteristics of mouse strains. Pp. 1-18 in Development of the Rodent as a Model System of Aging, Book II, D. C. Gibson, R. C. Adelman, and C. Finch, eds. DHEW Pub. No. (NIH) 79-161. Washington, D.C.: U.S. Department of Health, Education, and Welfare.

Albert, T. F., A. L. Ingling, and J. N. Sexton. 1976. Permanent outdoor housing for woodchucks, *Marmota monax*. Lab. Anim. Sci. 26:415-418.

Altman, P.L., and D.D. Katz, eds. 1979a. Inbred and Genetically Defined Strains of Laboratory Animals. Part II: Hamster, Guinea Pig, Rabbit, and Chicken. Bethesda, Md.: Federation of American Societies for Experimental Biology. 418 pp.

Altman, P.L., and D.D. Katz, eds. 1979b. Inbred and Genetically Defined Strains of Laboratory Animals. Part II: Hamster Guinea Pig, Rabbit, and Chicken. Bethesda, Md.: Federation of American Societies for Experimental Biology. 319 pp.

Altman, P. L. 1985. Pathology of Laboratory Mice and Rats. McLean, Va.: Federation of American Societies for Experimental Biology and Pergamon Infoline.

Andervont, H. B., and T. B. Dunn. 1962. Occurrence of tumors in wild house mice. J. Natl. Cancer Inst. 28:1153-1163.

Araki, K., J.-I. Miyazaki, O. Hino, N. Tomita, O. Chisaka, K. Matsubara, and K.-I. Yamamura. 1989. Expression and replication of hepatitis B virus genome in transgenic mice. Proc. Natl. Acad. Sci. USA 86:207-211.

Benjamin, S. A., and A. L. Brooks. 1977. Spontaneous lesions in Chinese hamsters. Vet. Pathol. 14:449-462.

Boice, R. 1971. Laboratizing the wild rat (*Rattus norvegicus*). Behav. Meth. Res. Instru. 3:177-182.

Bonhomme, F., and J. L. Guénet. 1989. The wild house mouse and its relatives. Pp. 649-662 in Genetic Variants and Strains of the Laboratory Mouse, 2d ed., M. F. Lyon, and A. G. Searle, eds. Oxford: Oxford University Press.

Bronson, F. H., and B. E. Eleftheriou. 1963. Influence of strange males on implantation in the deermouse. Gen. Comp. Endocrinol. 3:515-518.

Bronson, R. T. 1990. Rate of occurrence of lesions in 20 inbred and hybrid genotypes of rats and mice sacrificed at 6 month intervals during the first year of life. Pp. 279-358 in Genetic Effects on Aging, 2nd ed., D. E. Harrison, ed. Caldwell, N.J.: Telford Press.

Brooks, A. L., D. K. Mead, and R. F. Peters. 1973. Effect of aging on the frequency of metaphase chromosome aberrations in the liver of the Chinese hamster. J. Gerontol. 28:452-454.

Burek, J. D. 1978. Pathology of Aging Rats: A morphological and experimental study of teh age-associated lesions in aging BN/Bi, WAG/Rij, and (WAG x BN) Fob 10s rats. West Palm Beach, Fla.: CRC Press. 230 pp.

Burstein, D., J. P. Mordes, D. L. Greiner, D. Stein, N. Nakamura, E. S. Handler, and A. A. Rossini. 1989. Prevention of diabetes in BB/Wor rat by single transfusion of spleen cells; parameters that affect degree of protection. Diabetes 38:24-30.

CCAC (Canadian Council on Animal Care). 1984. Wild vertebrates in the field and in the laboratory. Pp. 191-204 in Guide to the Care and Use of Experimental Animals, Vol. 2. Ottawa, Ontario: Canadian Council on Animal Care.

Chappel, C. I., and W. R. Chappel. 1983. The discovery and development of thee BB rat colony: an animal model of spontaneous diabetes mellitus. Metabolism 32(suppl. 1):8-10.

Cheal, M. L. 1986. The gerbil: a unique model for research on aging. Exp. Aging Res. 12:3-21.

Clark, J. D. 1984. Biology and diseases of other rodents. Pp. 183-205 in Laboratory Animal Medicine, J. G. Fox, B. J. Cohen, and F. M. Loew, eds. Orlando, Fla: Academic Press.

Clough, G. 1991. Suggested guidelines for the housing and husbandry of rodents for aging studies. Neurobiol. Aging 12:653-658.

Coleman, D. L., J. E. Kuzava, and E. H. Leiter. 1990. Effect of diet on the incidence of diabetes in non-obese diabetic (NOD) mice. Diabetes 39:432-436.

Deamond, S. F., and S. A. Bruce. 1991. Age-related differences in promoter-induced extension of in vitro proliferative life span of Syrian hamster fibroblasts. Mech. Aging Dev. 60:143-152.

Deamond, S. F., L. G. Portnoy, J. D. Strandberg, and S. A. Bruce. 1990. Longevity and age-related pathology of LVG outbred golden Syrian hamsters (*Mesocricetus auratus*). Exp. Gerontol. 25:433-446.

Démant, P., and A. A. M. Hart. 1986. Recombinant congenic strains—A new tool for analyzing genetic traits determined by more than one gene. Immunogenetics 24:416-422.

Dewsbury, D. A. 1974a. The use of muroid rodents in the psychology laboratory. Behav. Meth. Res. Instru. 6:301-308.

Dewsbury, D. A. 1974b. Copulatory behaviour of white-throated wood rats (*Neotoma albigula*) and golden mice (*Ochrotomys nuttalli*). Anim. Behav. 22:601-610.

Dewsbury, D. A. 1984. Muroid rodents as research animals. ILAR News 28(1):8-15.

Dewsbury, D. A., and W. D. Dawson. 1979. African four-striped grass mice (*Rhabdomys pumilio*), a diurnal-crepuscular muroid rodent in the behavioral laboratory. Behav. Meth. Res. Instru. 11:329-333.

Doetschman, T., R. G. Gregg, N. Maeda, M. L. Hooper, D. W. Melton, S. Thompson, and O. Smithies. 1987. Targeted correction of a mutant HPRT gene in mouse embryonic stem cells. Nature 330:576-578.

Ediger, R. D. 1976. Care and management. Pp. 5-12 in The Biology of the Guinea Pig, J. E. Wagner and P. J. Manning, eds. New York: Academic Press.

Eisenberg, J. F. 1976. The heteromyid rodents. Pp. 293-297 in The UFAW Handbook on the Care and Management of Laboratory Animals, 5th ed., Universities Federation for Animal Welfare, eds. Edinburgh: Churchill Livingstone.

Festing, M. F. W. 1993. International Index of Laboratory Animals, 6th ed. Leicester, U.K. M. F. W. Festing. 238 pp. Available from M. F. W. Festing, PO Box 301, Leicester LE1 7RE, UK.

Festing, M. F. W., and D. D. Greenhouse. 1992. Abbreviated list of inbred strains of rats. Rat News Letter 26:10-22.

Fidler, I. J. 1977. Depression of macrophages in mice drinking hyperchlorinated water. Nature (London) 270:735-736.

Fine, J., F. W. Quimby, and D. D. Greenhouse. 1986. Annotated bibliography on uncommonly used laboratory animals: Mammals. ILAR News 29(4)1A-38A.

Gordon, J. W. 1993. Production of transgenic mice. Methods Enzymol. 225:747-770.

Gordon, J. W., G. A. Scangos, D. J. Plotkin, J. A. Barbosa, and F. H. Ruddle. 1980. Genetic transformation of mouse embryos by microinjection of purified DNA. Proc. Natl. Acad. Sci. USA 77:7380-7384.

Green, E. L. 1981. Mating systems. Pp. 114-185 in Genetics and Probability in Animal Breeding Experiments: a primer and reference book on probability, segregation, assortment, linkage and mating systems for biomedical scientists who breed and use genetically defined laboratory animals for research. London: Macmillan Press.

Guberski, D. L. 1993. Diabetes-prone and diabetes-resistant BB rats: Animal models of spontaneous and virally induced diabetes mellitus, lymphocytic thyroiditis, and collagen-induced arthritis. ILAR News 35:29-36.

Hall, J. E., W. J. White, and C. M. Lang. 1980. Acidification of drinking water: Its effects on selected biologic phenomena in male mice. Lab. Anim. Sci. 30:643-651.

Hansen, C. T., S. Potkay, W. T. Watson, and R. A. Whitney, Jr. 1981. NIH Rodents: 1980 Catalogue. NIH Pub. No. 81-606. Washington, D.C.: U.S. Department of Health and Human Services. 253 pp.

Harkness, J. E., and J. E. Wagner. 1989. The Biology and Medicine of Rabbits and Rodents, 3rd ed. Philadelphia: Lea & Febiger. 230 pp.

Harman, D. 1986. Free radical theory of aging: Role of free radicals in the origination and evolution of life, aging, and disease processes. Pp. 3-49 in Free Radicals, Aging, and Degenerative Diseases, J. E. Johnson, R. Walford, D. Harman, and J. Miguel, eds. New York: Alan R. Liss.

Hazzard, D. G., and J. Soban. 1989. Studies of aging using genetically defined rodents: A bibliography. Growth Dev. Aging 53:59-81.

Hazzard, D. G., and J. Soban. 1991. Addendum to: Studies of aging using defined rodents, a bibliography. Exp. Aging Res. 17:53-61.

Hedrich, H. J., and M. Adams, ed. 1990. Genetic Monitoring of Inbred Strains of Rats: A Manual on Colony Management, Basic Monitoring Techniques, and Genetic Variants of the Laboratory Rat. Stuttgart: Gustav Fischer Verlag. 539 pp.

Hermann, L. M., W. J. White, and C. M. Lang. 1982. Prolonged exposure to acid, chlorine, or tetracycline in drinking water: Effects on delayed-type hypersensitivity, hemagglutination titers, and reticuloendothelial clearance rates in mice. Lab. Anim. Sci. 32:603-608.

Hoffman, H. J. 1979. Survival distributions for selected laboratory rat strains and stocks. Pp. 19-34 in Development of the Rodent as a Model System of Aging, Book II, D. C. Gibson, R. C. Adelman, and C. Finch, eds. DHEW Pub. No. (NIH) 79-161. Washington, D.C.: U.S. Department of Health, Education, and Welfare.

Holloszy, J. O., and K. B. Schechtman. 1991. Interaction between exercise and food restriction: Effects on longevity of male rats. J. Appl. Physiol. 70:1529-1535.

Iwasaki, K., C. A. Gleiser, E. J. Masoro, C. A. McMahan, E. Seo, and B. P. Yu. 1988. The influence of dietary protein source on longevity and age-related disease processes of Fischer rats. J. Gerontol. 43:B5-B12.

Jaenisch, R. 1976. Germ line integration and Mendelian transmission of the exogenous Moloney leukemia virus. Proc. Natl. Acad. Sci. USA 73:1260-1264.

Kunst'yr, I., and S. Naumann. 1984. A contribution to guinea pig longevity data: Nine and one-half year-old guinea pig. Short communication. Z. Versuchstierkd. 26:57-59.

Leiter, E. H. 1990. The role of environmental factors in modulating insulin dependent diabetes. Pp. 39-55 in Current Topics in Immunology and Microbiology: The Role of Microorganisms in Non-infectious Disease, R. d.Vries, I. Cohen, and J. J. v. Rood, eds. Berlin: Springer Verlag.

Leiter, E. H. 1993. The NOD mouse: A model for analyzing the interplay between heredity and environment in development of autoimmune disease. ILAR News 35:4-13.

Lindsey, J. R. 1986. Prevalence of viral and mycoplasmal infections in laboratory rodents. Pp. 803-808 in Viral and Mycoplasmal Infectious of Laboratory Rodents: Effects on Biomedical Research, P. N. Bhatt, R. O. Jacoby, H. C. Morse III, and A. E. New, eds. Orlando, Fla.: Academic Press.

Lyman, C. P., R. C. O'Brien, G. C. Greene, and E. D. Papafrangos. 1981. Hibernation and longevity in the Turkish hamster *Mesocricetus brandti*. Science 212:668-670.

Makino, S., K. Kunimoto, Y. Muraoka, Y. Mizushima, K. Katagiri, and Y. Tochino. 1980. Breeding of a non-obese, diabetic strain of mice. Exp. Anim. 29:1-8.

Mansour, S. L., K. R. Thomas, and M. R. Capecchi. 1988. Disruption of the proto-oncogene int-2 in mouse embryo-derived stem cells: A general strategy for targeting mutations to non-selectable genes. Nature 336:348-352.

Marks, S. C., Jr. 1987. Osteopetrosis—Multiple pathways for the interception of osteoclast function. Appl. Pathol. 5:172-183.

Masoro, E. J. 1990. Animal models in aging research. Pp. 72-94 in Handbook of the Biology of Aging, 3rd ed., E. L. Schneider and J. W. Rowe, eds. New York: Academic Press.

Masoro, E. J. 1991. Use of rodents as models for the study of normal aging: conceptual and practical issues. Neurobiol. Aging 12:639-643.

McCormack, J. E., and A. L. Nutall. 1976. Auditory research. Pp. 281-303 in The Biology of the Guinea Pig, J. E. Wagner and P. J. Manning, eds. New York: Academic Press.

McPherson, C. W. 1963. Reduction of *Pseudomonas aeruginosa* and coliform bacteria in mouse drinking water following treatment with hydrochloric acid or chloring. Lab. Anim. Care 13:737-744.

Menich, S. R., and A. Baron. 1984. Social housing of rats: Life-span effects on reaction time, exploration, weight and longevity. Exp. Aging Res. 10:95-100.

Miller, R. A. 1991. Aging and immune function. Int. Rev. Cytol. 124:187-215.

Myers, D. D. 1978. Review of disease patterns and life span in aging mice: Genetic and environmental interactions. Birth Defects, Orig. Artic. Ser. 14:41-53.

Naji, A., W. K. Silvers, D. Bellgrau, and C. F. Barker. 1981. Spontaneous diabetes in rats: Destruction of islets is prevented by immunological tolerance. Science 213:1390-1392.

NRC (National Research Council), Institute of Laboratory Animal Resources, Committee on Immunologically Compromised Rodents. 1989. Immunodeficient Rodents: A Guide to Their Immunobiology, Husbandry, and Use. Washington, D.C.: National Academy Press. 246 pp.

NRC (National Research Council), Institute of Laboratory Animal Resources, Committee on Infectious Diseases of Mice and Rats. 1991. Infectious Diseases of Mice and Rats. Washington, D.C.: National Academy Press. 397 pp.

NRC (National Research Council), Institute of Laboratory Animal Resources Committee on Transgenic Nomenclature. 1993 Standardized nomenclature for transgenic animals. ILAR News 324(4):45-52.

Ohneda, A., T. Kobayashi, J. Nihei, Y. Tochino, H. Kanaya, and S. Makino. 1984. Insulin and glucagon in spontaneously diabetic non-obese mice. Diabetologia 27:460-463.

Oldstone, M. B. 1988. Prevention of type 1 diabetes in nonobese diabetic mice by virus infection. Science 239:500-502.

Pierpaoli, W., and N. O. Besedovsky. 1975. Role of the thymus in programming of neuroendocrine functions. Clin. Exp. Immunol. 20:323-338.

Potter, M. 1987. Listing of stocks and strains of mice in the genus *Mus* derived from the feral state. Pp. 373-395 in The Wild Mouse in Immunology, M. Potter, J. H. Nadeau, and M. P. Cancro, eds. Vol. 127 of Current Topics in Microbiology and Immunology. Berlin: Springer-Verlag.

Potter, M., J. H. Nadeau, and M. P. Cancro. 1986. The Wild Mouse in Immunology. Current Topics in Microbiology and Immunology, Vol. 127. New York: Springer Verlag. 395 pp.

Powles, M. A., D. C. McFadden, L. A. Pittarelli, and D. M. Schmatz. 1992. Mouse model *Pneumocystis carinii* pneumonia that uses natural transmission to initiate infection. Infect. Immun. 60:1397-1400.

Prochazka, M., E. H. Leiter, D. V. Serreze, and D. L. Coleman. 1987. Three recessive loci required for insulin-dependent diabetes in nonobese diabetic mice. Science 237:286-289.

Redfern, R., and F. P. Rowe. 1976. Pp. 218-228 in The UFAW Handbook on the Care and Management of Laboratory Animals, 5th ed, Universities Federation for Animal Welfare, eds. Edinburgh: Churchill Livingstone.

Reed, N. D., and J. W. Jutila. 1972. Immune responses of congenitally thymusless mice to heterologous erythrocytes. Proc. Soc. Exp. Bio. Med. 139:1234-1237.

Rossini, A. A., D. Faustman, B. A. Woda, A. A. Like, I. Szymanski, and J. P. Mordes. 1984. Lymphocyte transfusions prevent diabetes in the BioBreeding/Worcester rat. J. Clin. Invest. 74:39-46.

Rowlands, I. W., and B. J. Weir, 1974. The biology of hystricomorph rodents: the proceedings of a symposium held at the Zoological Society of London on 7 and 8 June, 1973. London. Published for the Zoological Society of London by Academy Press. 482 pp.

Rust, J. H., R. J. Robertson, E. F. Staffeldt, G. A. Sacher, D. Grahn, and R. J. M. Fry. 1966. Effects of lifetime periodic gamma-ray exposure on the survival and pathology of guinea pigs. Pp. 217-244 in Radiation and Aging. Proceedings of a colloquium held June 23-24, 1966, in Semmering, Austria. London: Taylor and Francis, Ltd.

Sacher, G. 1977. Life table modification and life prolongation. Pp. 582-638 in Handbook of the Biology of Aging, C. E. Finch, and L. Hayflick, eds. New York: Van Nostrand Reinhold.

Sacher, G. A., and R. W. Hart. 1978. Longevity, aging and comparative cellular and molecular biology of the house mouse, *Mus musculus*, and the white-footed mouse, *Peromyscus leucopus*. Birth Defects, Orig. Artic. Ser. 14:71-96.

Sadelain, M. W. J., H.-Y. Qin, J. Lauzon, and B. Singh. 1990. Prevention of type 1 diabetes in NOD mice by adjuvant immunotherapy. Diabetes 39:583-589.

Sage, R. D. 1981. Wild mice. Pp. 37-90 in The Mouse in Biomedical Research. Vol. I: History, Genetics, and Wild Mice, H. L. Foster, J. D. Small, and J. G. Fox, eds. New York: Academic Press.

Schneider, H. A. 1946. On breeding "wild" house mice in the laboratory. Proc. Soc. Exp. Biol. Med. 63:161-165.

Skalicky, M., H. Bubna-Littitz, and G. Hofecker. 1984. The influence of persistent crowding on the age changes of behavioral parameters and survival characteristics of rats. Mech. Aging Dev. 28:325-336.

Snyder, R. L. 1985. The laboratory woodchuck. Lab Anim. 14(1):20-32.

Soulez, B., F. Palluault, J. Y. Cesbron, E. Dei-Cas, A. Capron, and D. Camus. 1991. Introduction of *Pneumocystis carinii* in a colony of SCID mice. J. Protozool. 38:123-125S.

Sugawara, O., M. Oshimura, M. Koi, L. A. Annab, and J. C. Barrett. 1990. Induction of cellular senescence in immortalized cells by human chromosome 1. Science 247:707-710.

Takeda, T., M. Hosokawa, S. Takeshita, M. Irino, K. Higuchi, T. Matsushita, Y. Tomita, K. Yasuhira, K. Shimizu, M. Ishii, and J. Yamamuro. 1981. A new murine model of accelerated senescence. Mech. Aging Dev. 17:183-194.

Taketo, M., A. C. Schroeder, L. E. Mobraaten, K. B. Gunning, G. Hanten, R. R. Fox, T. H. Roderick, C. L. Stewart, F. Lilly, C. T. Hansen, and P. A. Overbeek. 1991. FVB/N: An inbred mouse strain preferable for transgenic analysis. Proc. Natl. Acad. Sci. USA 88:2065-2069.

Toneguzzo, F., A. C. Hayday, and A. Keating. 1986. Electric field-mediated DNA transfer: Transient and stable gene expression in human and mouse lymphoid cells. Mol. Cell. Biol. 6:703-706.

Umezawa, M., K. Hanada, H. Naiki, W. H. Chen, M. Hosokawa, M Hosono, T. Hosokaww, and T. Takeda. 1990. Effects of dietary restriction on age-related immune dysfunction in the senescence acclerated mouse (SAM). J. Nutr. 120:1393-1400.

van Abeelen, J. H., C. J. Janssens, W. E. Crusio, and W. A. Lemmens. 1989. Y-chromosomal effects on discrimination learning and hippocampal asymmetry in mice. Behav. Genet. 19:543-549.

van Zutphen, L. F. M., den Bieman, A. Lankhorst, and P. Démant. 1991. Segregation of genes from donor strain during the production of recombinant congenic strains. Lab. Anim. (London) 25:193-197.

Weihe, W. H. 1984. The thermoregulation of the nude mouse. Pp. 140-144 in Immune Deficient Animals, B. Sordat, ed. Basel: Karger.

Weir, B. J. 1967. The care and management of laboratory hystricomorph rodents. Lab. Anim. (London) 1:95-104.

Weir, B. J. 1976. Laboratory hystricomorph rodents other than the guinea-pig and chinchilla. Pp. 284-292 in The UFAW Handbook on the Care and Manageement of Laboratory Animals, 5th ed, Universities Federation for Animal Welfare, eds. Edinburgh: Churchill Livingstone.

Wilberz, S., H. J. Partke, F. Dagnaes-Hansen, and L. Herberg. 1991. Persistent MHV (mouse hepatitis virus) infection reduces the incidence of diabetes mellitus in non-obese diabetic mice. Diabetologia 34:2-5.

Wolf, N. S., W. E. Giddens, and G. M. Martin. 1988. Life table analysis and pathologic observations in male mice of a long-lived hybrid strain (A_f x C57BL/6)F_1. J. Gerontol. 43:B71-B78.

Young, R. A., and E. A. H. Sims. 1979. The woodchuck, *Marmota monax*, as a laboratory animal. Lab. Anim. Sci. 29:770-780.

Zitnik, G. D., S. A. Bingel, S. M. Sumi, and G. M. Martin. 1992. Survival curves, reproductive life span and age-related pathology of *Mus caroli*. Lab. Anim. Sci. 42(2):119-126.

Appendix

Sources of Information on Importing Rodents

Information on All Categories of Rodents

U.S. Department of Agriculture
Animal and Plant Health Inspection Service
Veterinary Services, Import/Export Products
Federal Building 22, Room 756
Hyattsville, MD 20782
Telephone: 301-436-7885

Information on Wild Rodents

U.S. Department of the Interior
Fish and Wildlife Service

Contact at one of the following addresses

New York, New York

700 Rockaway Turnpike
Lawrence, NY 11559
718-553-1767

Baltimore, Maryland

40 South Gay Street, Room 405
Baltimore, MD 21202
410-962-7980

Los Angeles, California

370 Amapola Avenue, Room 114
Torrance, CA 90501
310-297-0063

San Francisco, California

1633 Bayshore Highway, Suite 248
Burlingame, CA 94010
415-876-9078

Miami, Florida

10426 NW 31st Terrace
Miami, FL 33172
305-526-2789

Honolulu, Hawaii

PO Box 50223
Honolulu, HI
808-541-2681

Chicago, Illinois

10600 Higgens Road, Suite 200
Rosemont, IL 60018
708-298-3250

New Orleans, Louisiana

2424 Edenborn Road, Room 100
Metairie, LA 70001
504-589-4956

Seattle, Washington

121 107th NE, Suite 127
Bellevue, WA 98004
206-553-5543

Dallas/Fort Worth, Texas

PO Box 610069
D/FW Airport, TX 75261-0069
214-574-3254

Portland, Oregon

9025 SW Hillman Court, Suite 3134
Wilsonville, OR 97070
503-682-6131

For Customs Regulations

U.S. Department of the Treasury
U.S. Customs Service
(For local office, check lisings in telephone directory.)

Index

A

B

C

D

E

F

G

H

I

K

L

M

S

T

U

V

W

Z